Controlling
Human Heredity

THE CONTROL OF NATURE
Series Editors:
Margaret C. Jacob and Spencer R. Weart

PUBLISHED
SCIENTISTS AND THE DEVELOPMENT OF NUCLEAR WEAPONS
From Fission to the Limited Test Ban Treaty, 1939-1963
Lawrence Badash

EINSTEIN AND OUR WORLD
David Cassidy

NEWTON AND THE CULTURE OF NEWTONIANISM
Betty Jo Teeter Dobbs and Margaret C. Jacob

CONTROLLING HUMAN HEREDITY
1865 to the Present
Diane B. Paul

FORTHCOMING
SCIENCE AND TECHNOLOGY IN THE INDUSTRIAL REVOLUTION
Eric Brose

GENDER AND SCIENCE
Paula Findlen and Michael Dietrich

THE SCIENTIFIC REVOLUTION FROM COPERNICUS TO NEWTON
James Jacob

SCIENCE AND TECHNOLOGY IN TOTALITARIAN STATES
Paul Josephson

CONTROL OF NATURE

C
O
N
T
R
O
L

O
F

N
A
T
U
R
E

Controlling
Human Heredity

1865 to the Present

Diane B. Paul

HUMANITIES PRESS
NEW JERSEY

First published in 1995 by Humanities Press International, Inc.
165 First Avenue, Atlantic Highlands, New Jersey 07716

Library of Congress Cataloging-in-Publication Data
Paul, Diane B., 1946–
    Controlling human heredity : 1865 to the present / Diane B. Paul.
      p. cm. — (Control of nature)
    Includes bibliographical references and index.
    ISBN 0–391–03915–6 (cloth). — ISBN 0–391–03916–4 (pbk.)
    1. Eugenics—History. 2. Genetic engineering—Social aspects.
  3. Human reproductive technology—Social aspects. I. Title.
  II. Series.
  HQ751.P38 1995
  363.9'2'09—dc20   95–12762
                      CIP

A catalog record for this book is available from the British Library.

To the Memory of
George Armstrong Kelly:
teacher, scholar, mentor, friend

Contents

List of Illustrations

Series Editors' Preface

THIS SERIES OF historical studies aims to enrich the understanding of the role that science and technology have played in the history of Western civilization and culture and, through that, in the emerging world civilization. Each author has written with students and general readers, not specialists, in mind, and the volumes, of which this is the fourth to appear, have been written by scholars distinguished in their particular fields. The author of this book, Diane B. Paul, a professor of political science at the University of Massachusetts and a research associate at Harvard University, has pioneered the study of the political issues raised by genetics. Uniquely she has brought historical and feminist perspectives to the ethical and political questions posed by the emerging power of medical science to engage in genetic engineering.

This book on the science and politics of human heredity does not aim only to lay out some basic historical information, which could only be a sample of the many complex developments that scholars are exploring. Still more this volume intends to show the chief questions and debates that inform the current historical scholarship.

The current debates as presented here, in this instance on genetics, emphasize the "Control of Nature." While not excluding a discussion of how knowledge itself develops, how it is produced through the interplay of research into nature with the values and beliefs of the researcher, this volume—like all the others in the series—looks primarily to how science and technology interact with economic, social, political, and intellectual life, in ways that transform the relationship between human beings and nature. In every volume we are asking the student to think about how the modern world came to be invented, a world where the call for progress and the need to respect humanity and nature produce a tension, on the one hand liberating, on the other threatening to overwhelm human resources and ingenuity. The scientists whom you will meet here could not in every case have foreseen the kind of power that modern science and technology now offer. But they were also dreamers and doers—as well as shrewd promoters—who changed forever the way people view the natural world.

MARGARET C. JACOB
SPENCER WEART

Acknowledgments

My first debt is to friends who set aside their own tasks to help me meet urgent deadlines. I was constantly sustained by their generosity, encouragement, and unwavering faith that the job would get done. I am particularly grateful to two colleagues at the University of Massachusetts at Boston: Edna Seaman and Louise Z. Smith of the Biology and English Departments respectively. I would still be writing this book (or perhaps given up) were it not for their professional and personal support, which ranged from commenting on draft chapters to assuming household chores. For their friendship and sorely-tried patience, I am also indebted to the other members of the "birthday group"— Joan, Rachel, Kate, Linda, and Julie—and to Evelyn Fox Keller.

Twenty years ago I made a radical shift in the direction of my research. Hired to teach urban policy and political theory, I announced to my department that I intended to study biology in order to research the political dimensions of evolutionary theory and genetics. My colleagues in Political Science were as supportive as they were stunned by this decision. It has been my good fortune, rare in academia, to be a member of a department that is also a tolerant and nurturing community.

This book developed in part out of the experience of teaching a course on "The Darwinian Revolution" to a wide array of students— from senior Biology majors knowing little social history to freshman History majors knowing virtually nothing of evolution. Their struggles to cope with unfamiliar concepts spurred me to rethink and clarify my lectures. This book was framed with their questions, enthusiasms, and interests in view.

I owe special thanks to Jon Marks, Leila Zenderland, Robert Resta, Dan Wikler, and especially Margaret Jacob for insightful comments on various chapters, to Tom McMullin for advice on sources, to Gar Allen, Ute Deichmann, and the staff of the American Philosophical Society for help in locating illustrations, and to Pat Wright of Cambridge Wordwright for his superb editing.

My research on post-World War II developments in human genetics was supported by a grant from the Division of Research Programs of the National Endowment for the Humanities.

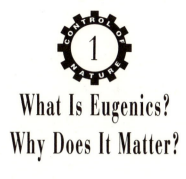

What Is Eugenics?
Why Does It Matter?

I N THE LAST two decades of the nineteenth century and the first three of this century, it was widely assumed that human mental, temperamental, and moral traits were determined by heredity. Shiftlessness, religiosity, courage, patriotism, a sense of humor, love of beauty, taste for philosophy, trustful nature, and a tendency to wander were only a few of the traits ascribed to good or bad blood. "Children inherit their minds and dispositions in the same way and to the same degree as they inherit their bodies," according to a popular book, *The Fruit of the Family Tree*. "Do you know," the author asked, "that good and bad housekeeping, good and bad citizenship, bright and dull minds, good and bad health, happiness and unhappiness are largely due to the sort of ancestors a man [*sic*] had, and that such things can be attained only to a limited extent by any economic 'system' or scheme of education?" (Wiggam 1924, 292).

That society ought to foster the breeding of those who possessed favorable traits ("positive" or "constructive" eugenics) and discourage or prevent the breeding of those who did not ("negative eugenics") seemed obviously to follow. Geneticists assured the public that the tools for the job were at hand. Samuel J. Holmes of the University of California told lay readers that anyone abreast of genetics and in control of human matings could, in a few generations, "breed a race of idiots, a race of dwarfs, a race of giants, an albino race, an insane race, a race of moral imbeciles, a race which would almost invariably get drunk in the presence of alcohol, a race of preeminent mental ability, or a race of unusual artistic talent." From a scientific perspective, the task is "easy," he claimed, and some control of breeding is badly needed. There is no excuse for allowing "degenerate human beings" to propagate their kind. He noted that some people counsel caution on the grounds that we lack consensus

1

on what kinds of people are desirable and enough knowledge about the laws of heredity to breed them. On both scores, Holmes thought them wrong. We want people who are healthy, good-natured, emotionally stable, sympathetic, and smart. We do not want idiots, imbeciles, paupers, and criminals (Holmes 1923, 200–203).

In the United States, Canada, Britain, and much of Europe, the concept of genetic perfectibility underlay some of the most sharply defined fissures in the intellectual and moral landscape. Holmes's views were widely shared by those who were white, Anglo-Saxon, Protestant, and middle class. To individuals who had achieved social success, controlled reproduction seemed only good common sense. Advocates and critics of capitalism, defenders and opponents of birth control, social revolutionaries and reactionaries lined up on the question of human breeding. How did eugenics come to exert such powerful and broad appeal? What events shaped its direction? Whose interests did it finally serve? Why did it fall into disrepute? In spite of eugenics' bad reputation, has it survived in other guises? Indeed, what is eugenics, and what relationship does it have to modern genetics? These are the questions this book aims to answer.

These questions have acquired a new urgency in light of developments in genetic medicine. Only a few years ago, it seemed clear that eugenics had been wholly discredited by its association with race and class prejudice, and in particular with the crimes of the Third Reich. The movement appeared to be dead. Or was it just sleeping? Some people have recently questioned whether the reaction to Nazi crimes produced more than a temporary hiatus in eugenic theory and practice (Neuhaus 1990, 1). They fear that eugenics is back—in the benevolent guise of medical genetics.

The recent revival of concerns about eugenics is not surprising. Twenty-five years ago, prenatal tests were rare. Although amniocentesis was developed in the 1960s, there was little demand for it until the Supreme Court legalized abortion. Since *Roe v. Wade* in 1973, the number of women undergoing prenatal diagnosis has continually grown, as have the kinds of fetal anomalies detected. Predictive tests are now also available for a number of late-onset diseases, such as Huntington's chorea. And "carrier" tests, which identify individuals whose offspring will be affected if they inherit the same gene from both parents, are expanding from very rare diseases to more common ones, such as cystic fibrosis, where about 5 percent of people of Northern European ancestry carry the gene. Thus more and more people are being offered (and may feel social pressure to accept) a rapidly increasing array of genetic tests (Lippman 1991a).

As a consequence, they are faced with novel and perplexing decisions. Predictive tests create the need to make choices. If I learn that my child will have Down syndrome, should I carry the fetus to term? What if we already have one child and want the second to be of the opposite sex? What if I carry the same gene for Tay-Sachs disease as does my prospective spouse? Should I marry someone else? What if I have inherited the dominant gene for Huntington's chorea? Should I avoid having children at all? To some, breeding decisions based on genetic tests appear to be eugenics in modern dress. Others insist that if decisions are voluntary, the label is wrongly applied. This dispute will not be easily resolved. As we will see, eugenics has always been a protean concept. Almost from the start, eugenics has meant different things to different people.

The first to define it was Francis Galton (1822–1911), a man of many talents who is best known for his pioneering work in statistics and its application to problems of human heredity. In 1865, Galton proposed that humans take charge of their own evolution, but it was not until 1883 that he invented a name for this program of selective breeding: *eugenics*, from the Greek *eugenes* for "good in birth." Galton defined eugenics broadly, as "the science of improving stock, which is by no means confined to questions of judicious mating, but which ... takes cognisance of all influences that tend in however remote degree to give the more suitable races or strains of blood a better chance of prevailing speedily over the less suitable than they otherwise would have had" (1883, 24). Galton later experimented with a variety of different formulations such as "the study of agencies under social control which may improve or impair the racial qualities of future generations" (1909c, 81) and "the science which deals with all influences that improve the inborn qualities of a race; also with those that develop them to the utmost advantage" (1909a, 35).

In all of these definitions, eugenics sounds rather innocuous. Indeed, most medical genetics would fall within its domain. Galton's category is broad enough to include prenatal diagnosis and other developments in genetic medicine. But today eugenics is generally assumed to be bad. No wonder that the new technologies have aroused great anxiety and that eugenics figures so prominently in discussions of social and ethical issues in biomedicine. Some people fear that our enhanced ability to make reproductive choices, or the prospect of direct alteration of particular genes, will usher in a "new eugenics." Others think it already has done so. Still others dismiss any relation at all between contemporary genetic medicine and eugenics. Who is right depends not just on the facts but on what is meant by eugenics. There may be near-consensus that eugenics has been discredited, but there is no consensus on what eugenics *is*.

Most contemporary definitions are much narrower than Galton's. The eugenic label is often restricted to policies that are coercive. These definitions create a sharp distinction between eugenics and medical genetics. Other definitions erase this distinction. A spate of recent books and articles has warned of eugenics as the unintended result of individual choices. On this view, the greatest danger arises not from coercion but its reverse: our enhanced ability to *choose* the kind of children we want (Duster 1990; Wright 1990).

However defined, the word *eugenics* carries ominous connotations. It packs a powerful emotional punch. To characterize medical genetics as eugenics is usually to condemn it. The obligatory accounts of eugenics in discussions of social issues in genetic medicine are meant to warn against misuses of the new technologies; what they say about eugenics is thus both damning and dramatic. For this reason, few contemporary geneticists would want to claim Galton as forebear. Yet in the 1910s and 1920s, eugenics was simply considered applied human genetics. While we will see that the early geneticists divided on many specific issues, they were virtually united in their view that social ills such as poverty and crime resulted from defects in heredity. These conditions were not the fault of the pauper or prostitute or thief: one was no more responsible for inheriting character traits than eye color. The problem of "racial degeneration" was thus viewed in new and fashionably scientific terms rather than in traditionally moral ones. Eugenicists sought the key to understanding degeneracy in the new science of genetics, and its solution in the practical application of its principles. Both understanding and control were considered matters of great urgency, since the problem of degeneration appeared to be rapidly growing worse. Civilized societies now kept alive the physically and mentally weak. Worse, the defectives were reproducing at a much faster rate than were the strong and smart.

Galton was the first to signal the alarm. From studies of eminent British families, he concluded that mental and moral traits such as scientific and artistic ability, veracity, courage, a courteous disposition, and consistency of purpose were passed from parents to offspring. Some individuals were innately talented, vigorous, and wise, whereas others were born weak in mind and body. "Whether it be in character, disposition, energy, intellect, or physical power," he maintained, "we each receive at our birth a definite endowment, allegorised by the parable related in St. Matthew, some receiving many talents, others few" (1909b, 3).

Alas, the worthiest individuals seemed to leave the fewest offspring. Given the complexity of modern life, Galton warned, this trend could be catastrophic. But a remedy was fortunately at hand. Those highest in civic worth should be encouraged to have more children; the stupid and improvident, fewer or none. In Galton's view, it was inconsistent to improve varieties of domestic animals while leaving human heredity to chance. He thus proposed extending the stockbreeders' methods to his own species. "If a twentieth part of the cost and pains were spent in measures for the improvement of the human race that is spent on the improvement of the breed of horses and cattle, what a galaxy of genius might we not create!" he exclaimed (1865, 165). While emphasizing the need to encourage reproduction by the fit, Galton also approved of negative measures. In his unpublished utopian fantasy "Kantsaywhere" (1910), citizens were allowed to reproduce according to quotas determined by their ranking on physical and mental tests. For the "Unfit" to breed at all would be looked on by the inhabitants as a crime against the state. Quota violations would be punished by penalties ranging from fines to lifelong segregation in labor colonies (Pearson 1930, 420).

Galton was not the first to suggest that matings might be controlled in the interest of improving the human race. In Plato's *Republic*, written in the fifth century B.C., rulers decided who would bear how many children, and imperfect offspring were hidden away. In the fifth book of the *Laws*, he elaborated the analogy between animal and human breeding. Just as shepherds and breeders purge their herds, Plato argued, so must the legislator purify the state. His work provided the inspiration for other eugenic utopias, such as Tomasso Campanella's *City of the Sun* (1623), a community in which unions were arranged by a "Great Master" (aided by chief matrons), who allowed only superior youths to procreate. In more recent times, John Humphrey Noyes, who in 1848 founded a Christian community in Oneida, New York, predicted that "the time will come when involuntary and random procreation will cease, and when scientific combination will be applied to human generation as freely and successfully as it is to . . . animals" (Carden 1969, 61). But the idea of controlled human breeding remained largely abstract until the nineteenth century, when it converged with modern statistical and evolutionary theory. Noyes only began to experiment with the practice he labeled "stirpiculture" after reading works by Darwin and Galton in the 1860s. "Blood tells" is an old folk maxim, but only in the nineteenth century could it be made at all precise. When joined to the principle of evolution by natural selection, proposed by Galton's cousin Charles Darwin (1809–1882), and to modern genetics, it was to prove explosive.

The specter of evolutionary degeneration haunted middle-class Victorians. In his 1859 book *On the Origin of Species*, Darwin introduced the concept of natural selection: a process in nature, akin to plant and animal breeding, whereby favorable variations are preserved and injurious ones destroyed. In Darwin's view, organisms with favorable variants were more likely to survive in the intense struggle for existence in nature. While Darwin declined to draw explicit lessons for his own species, many of his compatriots were quick to note that civilized societies interfered with this struggle in myriad ways and to express alarm at the consequences. They charged that humanitarian measures protected society's least capable members, who would once have succumbed to disease or starvation. Paupers and imbeciles now survived and bred their like, while the middle class exercised reproductive restraint. Thus it seemed that the beneficial effects of natural selection were checked or even reversed.

Darwin finally published his book on human evolution in 1871. *The Descent of Man* warned that if the lower classes continued to outbreed their social superiors, evolutionary regress would result. But as we will see in chapter 2, Darwin also rejected proposals to withdraw aid from the weak and helpless. Indeed, he proposed no definite solution but only called for more education in the principles of heredity. Darwin had written anxiously about differential rates of reproduction and had reinforced the analogy of human with animal breeding, thus pointing to both the problem and the solution that would seem obvious to many. But he did not take this step, nor did many of his contemporaries.

It was only after the turn of the century that the idea of controlled human breeding really caught on. Organized eugenics movements were first founded in Germany in 1904, in Britain in 1907, and in the United States about 1910 (Allen, in press). To the extent that eugenics involved interference with the hitherto private sphere of reproduction, it depended on a shift in attitudes toward the state. In 1868, the Scots essayist William Greg (whose work greatly influenced Darwin) imagined a republic in which "paternity should be the right and function exclusively of the elite of the nation." But Greg understood that his compatriots would reject the expansion of state power required to breed a better race. Like Darwin, he was thus left to trust in "the slow influences of enlightenment and moral susceptibility, percolating downwards and in time permeating all ranks" (Greg 1868, 361–62). The application of eugenics—for example, as in programs of eugenical sterilization—awaited the rise of the welfare state.

As we will see in chapter 3, a shift in attitudes toward heredity also helped turn abstract idea into practical program. Darwin and nearly all his contemporaries believed that traits acquired by organisms were transmissible to their progeny (the principle of "inheritance of acquired characteristics"). In this view, positive changes in environment should prompt improvements in heredity. This view of heredity was generally associated with positive attitudes toward social reform. But it was challenged in the late nineteenth century by August Weismann, who argued that the hereditary material was impervious to changes in environment. Weismann's argument was reinforced, in turn, by the work of Gregor Mendel, which first came to public notice in 1900. As a consequence, many people came to think that the only way to improve the race was through selective breeding.

To Victorians living in cities overrun by displaced agricultural laborers and surplus factory hands, the problem of the "differential birthrate" had also come to seem urgent. Galton, Darwin, and Greg had expressed dismay at what they considered the profligate breeding of the poor in comparison with the middle class. In the early twentieth century, a raft of reports seemed to confirm the alarmists' worst fears about racial degeneration. The end of the Boer War (1899–1902) was followed by dramatic reports of the large number of recruits (over 40 percent in the city of Manchester) who had been declared physically unfit for military service (Kramnick and Sheerman 1993, 38). Moreover, numerous demographic studies demonstrated an inverse correlation between fertility and socioeconomic status, with the birthrate dropping much more steeply among the middle and upper middle classes than among workers and agricultural laborers (Soloway 1990, 10–17). In 1906, David Heron of the Galton Laboratory published a report on the fertility of married women in 27 London districts and concluded that "the morally and socially lowest classes" were producing the largest families, the most prosperous and cultured the smallest (quoted in Soloway 1990, 14). Ethel Elderton's *Report on the English Birth Rate* (1914) looked beyond London to reach a similar conclusion: that bad social conditions correlate with large families. The Fabian socialist Sidney Webb took a somewhat different tack. In *The Decline of the Birth-Rate* (1907), Webb argued that the problem was not fertility differences among classes but rather the low reproductive rates of the most prudent and disciplined members of every class. (He also lamented the high reproductive rates of Irish Roman Catholics, Poles, Russians, and Jews.) Whatever their attitude toward the working class generally, these researchers agreed that society's best were being swamped by its worst. Some feared that Britain's military readiness and capacity for exercising imperial power would be fatally undermined (Kramnick and Sheerman 1993, 38–39; Searle 1976, 20–23).

Worst of all was the apparently high fertility of the "feebleminded."

A series of American studies tracing the descendants of problem families claimed to demonstrate the prolific breeding of mental defectives. "In former times the numbers of the feebleminded were kept down by the stern processes of natural selection," wrote Lothrop Stoddard in *The Revolt against Civilization,* "but modern society and philanthropy have protected them and thus favored their rapid multiplication." Stoddard claimed that numerous scientific studies all told the same story: the feebleminded were "spreading like cancerous growths . . . infecting the blood of whole communities, and thriving on misguided efforts" to improve their social condition (1923, 94). Adding to the alarm in America was a massive influx of immigrants from southern and eastern Europe. As we will see in chapter 6, across the political spectrum these newcomers were viewed as a biological as well as economic and cultural threat. Russians, Hungarians, Poles, Italians, and Greeks were thought to be both disproportionately feebleminded and fertile. Their mating with old-stock Americans seemed certain to produce biological degeneration.

In both America and Britain, the official eugenics societies were small: the British Eugenics Education Society had approximately 1,200 members; the American Eugenics Society, no more than that (Kevles 1985, 59; Soloway 1990, 34). However, their members made up in influence what they lacked in numbers. Both societies were dominated by professionals such as professors, social workers, lawyers, doctors, teachers, and ministers but also included individuals with substantial business interests (and in Britain a smattering of aristocrats). The small memberships reflect the extreme dispersion of the movement, which especially in the United States was constituted by many geographically distant and diverse groups, rather than weak public support for the cause.

While the official societies had modest budgets, money flowed to other institutions and projects. Foundations and wealthy individual donors provided especially generous funding for the American movement. Philanthropic support for the Eugenics Record Office (ERO) at Cold Spring Harbor, New York, was particularly important. The ERO provided some coherence to a diffuse movement by acting as a clearinghouse for eugenics efforts, publishing the *Eugenical News,* organizing research, serving as a repository for data on the traits of hundreds of thousands of individuals, and operating a summer school for eugenic field workers (Allen 1989). In 1910, the geneticist Charles Davenport (1866–1944), director of the respected Station for the Experimental Study of Evolution at Cold Spring Harbor, convinced the widow of railroad magnate Edward Henry Harriman to fund an institution dedicated to improving human heredity. (Davenport had taught genetics to their daughter, Mary.) Mrs. Harriman paid for the buildings and, during the next seven years,

FIGURE 1.1 Banquet, Delegates to the National Conference on Race Betterment, Battle Creek, Michigan, January 10, 1914. From *Proceedings, First National Conference on Race Betterment*, 1914 (Francis Countway Medical Library, Harvard University).

contributed $246,000 toward the ERO's total operating costs of about $440,000. In 1917, when the Carnegie Institution of Washington, which operated the station, took over the responsibility for operating costs, she provided an additional endowment of $300,000 (Allen 1986, 235–36). Between 1918 and 1939, when the ERO closed its doors, the Carnegie Foundation provided about half a million dollars for operating expenses. John D. Rockefeller Jr., son of the founder of the Standard Oil Company, contributed additional funds for the training of eugenic field workers (262–63). A number of other foundations and private donors also provided substantial support to the eugenics movement. The Battle Creek Race Betterment Foundation, endowed by cereal king John H. Kellogg, sponsored a series of well-publicized eugenics conferences. The Human Betterment Foundation, founded by the millionaire banker and citrus grower Ezra Gosney (who had been inspired by the ERO), financed influential studies promoting eugenical sterilization (Reilly 1991, 80–81). Many smaller foundations and wealthy, often eccentric, individuals also helped fund particular institutes and underwrote research projects and publications.

These funds helped swell the flood of eugenics literature. Articles on eugenic themes appeared frequently in popular magazines and newspapers. By 1910, eugenics was one of the most frequently referenced topics in the *Reader's Guide to Periodical Literature* (Reilly 1991, 18). Books, public lectures, church services, and even films relayed the message that good breeding was everyone's business.

Both in Britain and in America, many Protestant clergy also helped spread the message (Jones 1986, 47–49; Soloway 1990, 80–85). In 1912, the Reverend W. T. Sumner, dean of the Episcopal Cathedral of Saints Peter and Paul in Chicago, announced that clergy in that cathedral would refuse to marry couples who could not present a physician's certificate showing that both parties were physically and mentally fit. Shortly thereafter, the dean's views were unanimously endorsed by the 200 clergy attending the meeting of the federated churches of Chicago. From the sampling of opinion published in the *New York Times* of July 12, 1912, it appears that clergy from a wide array of demoninations applauded Dean Sumner. "Mawkish sentiment must unquestionably yield to the high issues involved in eugenics," wrote one pastor from New York. Another asked rhetorically, "Have not the recent studies of the criminal, the pauperized, and the feeble-minded given us an abundance of testimony as to the awful cost which society pays for its failure to control the institution of marriage?"

In 1914, 44 American colleges and universities offered courses in eugenics; by 1928, the number had increased to 376 (Cravens 1978,

53). Eugenics was endorsed in over 90 percent of high school biology textbooks (Selden 1989). During the 1920s, "Fitter Families for Fitter Firesides" and "Better Babies" contests were featured at American state fairs, as well as exhibits illustrating the financial costs of leaving heredity to chance alongside livestock exhibits that implicitly demonstrated the value of selective breeding. Mary Watts, co-organizer of the first "Fitter Families" contest at the 1920 Kansas Free Fair, explained that, "when someone asks what it is all about, we say, 'while the stock judges are testing the Holsteins, Jerseys, and whitefaces in the stock pavilion, we are judging the Joneses, Smiths, & the Johnsons,' and nearly every one replies: 'I think it is about time people had a little of the attention that is given to animals'" (Rydell 1993, 49–50).

Of course there were countercurrents. Some even mocked the eugenics crusade. Thus, in Sinclair Lewis's 1925 novel *Arrowsmith*, the Eugenic Family at an Iowa fair is exposed as the criminal "Holton gang"; the parents turn out to be unmarried, and one of their five purported children suffers an epileptic fit during a health demonstration. Rejecting any interference with reproduction, the Catholic Church strongly opposed sterilization laws. In his 1930 encyclical "On Christian Marriage," Pope Pius XI repudiated eugenics as contrary to Church teachings on the sanctity of the family. Catholic philosophers and literary figures such as G. K. Chesterton (1874–1936) denounced it as part of "the modern craze for scientific officialism and strict social organization" (Chesterton 1922, vi). (Associating Catholicism with superstition and reaction, however, many Protestants felt confirmed in their belief that eugenics was rational and progressive.) Irrespective of religion, the claim that heredity determines character and explains social success conflicts with traditional moral maxims. Indeed, we will see in chapter 2 that Galton's open scorn for conventional pieties limited the appeal of his work. The claim that we are products of our genes is in particular tension with the quintessentially American view that we are products of education, experience, and initiative. Foreign observers often noted (sometimes with ridicule) the common American belief that people can pull themselves up by their own bootstraps. Eugenicists thus had to counter the powerful American faith in education and in the efficacy of moral effort. It is testimony to the strength of the social forces behind it that eugenics was able to flourish in the land of Horatio Alger.

Apart from the Catholic Church, Britain's Labour Party, and some liberal individualists, there was little moral opposition to eugenics before the Second World War, although many objections were raised to specific eugenic measures and to the racial, ethnic, and class biases they often reflected. Even harsh critics of the eugenics movement often shared

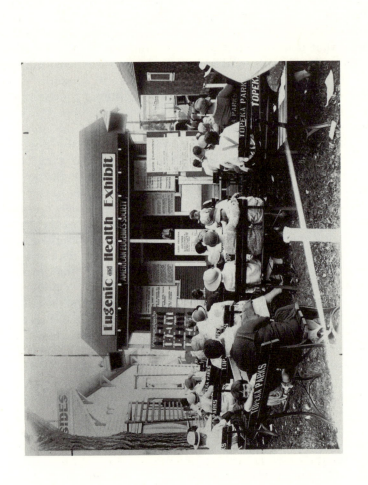

FIGURE 1.2 Eugenics Building, Kansas Free Fair, 1929. Courtesy of the American Philosophical Society, Philadelphia.

FIGURE 1.3   Flashing light sign used with small exhibits sponsored by the American Eugenics Society at many state fairs. Courtesy of the American Philosophical Society, Philadelphia.

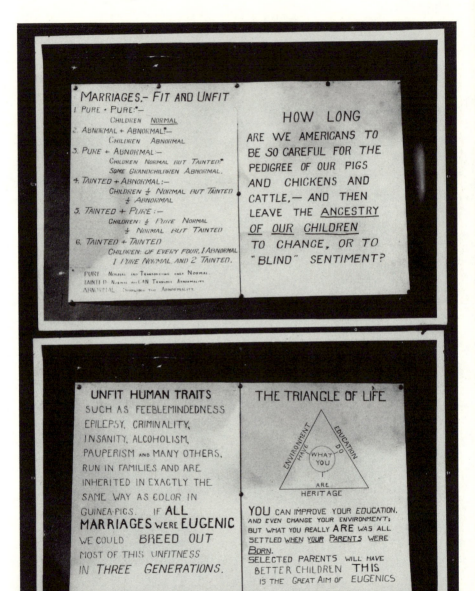

FIGURE 1.4   Charts used at Kansas Free Fair. Courtesy of the American Philosophical Society, Philadelphia.

FIGURE 1.5 Winner of trophy in Large Family Class, Kansas State Fair, 1925. Courtesy of the American Philosophical Society, Philadelphia.

FIGURE 1.6   Four generations at Fitter Families Contest, Kansas Free Fair, 1923. Courtesy of the American Philosophical Society, Philadelphia.

some of its assumptions. To denounce eugenics did not (and we will see in the last chapter even today does not) necessarily imply its wholesale rejection. The philosopher Leonard Hobhouse was a prominent critic of the British movement, but he believed that society should bar from breeding those "men and women who are not capable of independent existence but who continually drift to the gaol or the workhouse, who are fertile, and whose condition is asserted to be hereditary in a marked degree" (Hobhouse 1911, 45–46). In the United States, no academic is more associated with skepticism toward eugenics than the anthropologist Franz Boas. But in 1916 even he thought it proper to "suppress those defective classes whose deficiencies can be proved by rigid methods to be due to hereditary causes, and to prevent unions that will unavoidably lead to the birth of disease stricken progeny" (Boas 1916, 478). From a contemporary standpoint, the area of agreement between advocates and critics of eugenics was often surprisingly large. But these once-shared views are no longer considered respectable. Indeed, *eugenics* was converted in under fifty years from a term of praise to one of opprobrium.

Today, the images conjured up by *eugenics* range from ludicrous to loathsome. We think of photos of trophies awarded to eugenically superior families at American state fairs; of the legions of middle-class women, armed with "trait books," who judged at a glance the heredity of rural families; of the patients sterilized without their permission and sometimes even their knowledge; of the American army testing recruits' "native ability" with such questions as "The number of a Kaffir's legs is: 2, 4, 6, 8"; of Nazi persecution of Jews, homosexuals, and the "hereditarily sick." We thus think of eugenics as odious when not simply absurd. Those who espoused it were obviously villains and fools.

As we will see, elements in this picture are certainly true. Many eugenicists were racist and reactionary, even by the standards of their own times. For some of them, the individual counted for nothing, the larger community all; only the "fit" had a right to survive, or at least reproduce. One person who held this view was Madison Grant, a leading conservationist, president of the New York Zoological Society, and friend of Theodore Roosevelt, whose racial views he influenced. In *The Passing of the Great Race*, Grant wrote:

> Mistaken regard for what are believed to be divine laws and a sentimental belief in the sanctity of human life tend to prevent both the elimination of defective infants and the sterilization of such adults as are themselves of no value to the community. The laws of nature require the obliteration of the unfit, and human life is valuable only when it is of use to the community or race. (1916, 44–45)

Grant represents one end of a continuum of views in American eugenics before World War II. There were also eugenicists who did not share his political perspective. In some popular histories, eugenicists were all political reactionaries who held scientific views that were obviously preposterous. Indeed, it is often said that the views the eugenicists espoused had nothing to do with science. In this perspective, eugenics was based on a series of absurdly naive assumptions: that human mental traits were the products of single hereditary factors (after 1908 called "genes"), that individuals who possessed good and bad genes could be easily identified, and that preventing those with the bad genes from reproducing would rapidly reduce their numbers. The progress of science undermined and ultimately destroyed these illusions. Geneticists showed that most traits were generally influenced by multiple genes and that genes interacted with each other and with the environment in complex ways. Geneticists also demonstrated that most bad genes, being rare and recessive, would be hidden in apparently normal carriers. Since the afflicted themselves were but the tip of an iceberg, policies directed at them would have little effect. Eugenical segregation and sterilization were shown to be futile. The development of "real" genetics thus exposed the hollowness of eugenicists' claims. From this standpoint, genetics is science, eugenics a pseudo-science.

But this account makes a mystery of the fact that many distinguished geneticists promoted eugenics. Every member of the first editorial board of the American journal *Genetics*, founded in 1916, endorsed the movement (Ludmerer 1972, 34). In interwar Germany, where the science of human genetics was especially advanced, geneticists' support for eugenics was near universal. Some geneticists in North America, Britain, and Scandinavia did reproach "mainline" eugenicists for scientific naiveté and social bias. But these critics generally aimed to reform eugenics, not refute it. In Germany, criticism was almost wholly absent; it appears that not a single geneticist expressed reservations about the policies of the German movement (Harwood 1989, 262).

Charles Davenport was a respected scientist, who did important work both on the genetics of disease (such as Huntington's chorea) and normal human traits (such as skin and eye color). He also believed that "society must protect itself[;] as it claims the right to deprive the murderer of his life so also it may annihilate the hideous serpent of hopelessly vicious protoplasm" (Davenport 1909, 129). Edward M. East (1879–1938) was a Harvard geneticist who played a crucial role in convincing geneticists of the importance of Mendel's work. He also argued that black-white intermarriage was biologically ill advised, that immigration from southern and eastern Europe was a threat, and that public institutions were crowded with the genetically unfit. The German geneticist Eugen Fischer (1874–1967) was a key figure in the development of human genetics. In 1939 he said, "I reject Jewry with every means in my power, and without reserve, in order to preserve the hereditary endowment of my people" (quoted in Müller-Hill 1988, 12). These scientists, and many of their peers, held social views that most people today find repugnant. But they were not necessarily stupid. Some made important scientific contributions. Indeed, in some cases they made the discoveries that, in the conventional view, are said to have brought eugenics down. Furthermore, they were not all socially conservative.

We now know that eugenics was a more diverse movement, enjoying much broader political support, than one would imagine from conventional accounts, which also judge long-dead actors by the standards of our own (ostensibly enlightened) time. From many detailed and comparative studies, we have learned that eugenics took different forms in different countries (Adams 1990b; Allen, in press). Variation in religion, scientific tradition, and degree of racial mixing, among other factors, sometimes produced a moderate and hopeful cast to eugenics. Although there were about 30 movements worldwide, encompassing Latin America and Japan as well as America and Europe, generalizations have been based on a few cases. French and Brazilian eugenics, for example, varied in important respects from eugenics in the United States, Britain, and Germany (Schneider 1990, Stepan 1991). In the former countries, sterilization and other coercive measures were rejected. Racism was muted. Hereditary and environmental improvement were generally thought to go hand in hand.

The more humane movements show that very disparate policies were promoted under the eugenics banner. Indeed, these movements were so varied that their only commonality was a general concern with preventing biological degeneration. In France, physicians who warned prospective parents about the hazards of syphilis and alcohol termed their endeavors "eugenics." The recognition that a broad concern with biological improvement and decline could take different forms in different national contexts is important, for it alerts us to the potential plasticity of eugenics. But we would not be so alarmed about the prospects of a eugenics revival today if most eugenics movements had been compassionate. It is because these movements took other, more menacing forms that eugenics is a matter of such historical interest and such intense contemporary concern.

However, even in respect to eugenics in Germany, Russia, Scandinavia, Canada, and the two countries emphasized in this book—Britain and especially the United States—the conventional picture needs to be amended. It is true that most eugenicists, and in particular those associated with the organized eugenics societies, were politically conservative and socially pessimistic. But eugenics also appealed to a wide range of reformers. For example, in the state of Georgia, eugenical sterilization was promoted by the same New Dealers who worked for the provision of old-age pensions, free school books, and well-baby clinics. Their proposed sterilization law was vetoed—along with every item of New Deal legislation that passed his desk—by the reactionary white-supremacist Governor Eugene Talmadge. It finally passed in 1937 when Talmadge was replaced by a governor from the progressive wing of the Democratic Party (Larson 1991).

Eugenics even appealed to some whose politics were further to the left. In the 1920s, several Russian geneticists called for a "Bolshevik eugenics" based on Marxist principles. In their view, eugenics was a logical extension of the Marxist commitment to the scientific organization of society (Graham 1977). The distinguished geneticist Alexandr Serebrovsky claimed that if his country's population could be cleansed of hereditary defects, the time and money saved would make it possible to fulfill the Five-Year Plan in half the time (Adams 1990a, 180). Political radicals in other countries also endorsed eugenics. Alfred Ploetz, a key figure in the founding of the eugenics movement in Germany, was a socialist who lived (briefly) in a utopian commune in Iowa (Weiss 1990). Harold J. Laski, a leading theorist of British Fabian socialism, spoke for many in the movement when he warned in 1910 that "the different rates of fertility in the sound and pathological stocks point to a future swamping of the better by the worse. As a nation, we are faced by race suicide" (Laski 1910, 34). In America, where eugenics had a strong racist cast, a number of socialists likewise warned of race suicide. But as we will see in chapter 6, they usually had immigrants—especially from Asia—in mind.

The movement to legalize birth control was also strongly linked to eugenics. "*Laisser aller* in marriage is no wiser than in other parts of life," asserted the British birth control advocate Annie Besant (quoted in Bannister 1979, 174). Besant was another Fabian socialist, but as we will see in chapter 5, her view that the state had an interest in the biological quality of its future citizens was shared by mainstream feminists such as Marie Stopes in Britain and Margaret Sanger in the United States. Arguing that educated women, unlike the poor, already had access to birth control, they promised that reliable contraception would reduce births among paupers, criminals, and other undesirables. Sanger made the point succinctly in 1919: "More children from the fit, less from the unfit—that is the chief issue of birth control" (quoted in

Kennedy 1970, 115). Opponents of contraception, however, warned that the birthrates of college-educated women were already dangerously low and would decline still further if birth control was easily available.

In the late nineteenth and early twentieth centuries, eugenicists were found on every side of arguments about capitalism, war, and especially the role of women. Conservatives employed eugenic arguments to justify restrictions on birth control, suffrage, divorce, and women's educational and professional opportunities, while social radicals employed other eugenic arguments to assail them. Some eugenicists asserted that war would strengthen the race but more denounced it on the grounds that modern warfare sacrificed the healthiest and bravest men while sparing the physically and mentally unfit (Crook 1994, 83–91, 159–166). Most defended capitalism, whereas others argued that only in a classless society would it be possible to separate the genetic wheat from the chaff. Thus eugenicists were united only in their enthusiasm for technocratic solutions to social problems. We will fail to understand the appeal of eugenics to so many people with such divergent interests, training, and political orientations if we start with the assumption that it was patently absurd. We may find the wide enthusiasm for eugenics shocking, but it reflected scientific and social beliefs that our great-grandparents found satisfying and reasonable. These beliefs seemed just as obvious to them as our own convictions do to us. We need to understand why they believed what they did. It is also well to keep in mind that at least some contemporary taken-for-granted beliefs will undoubtedly appear as preposterous to our grandchildren as they would have to our forebears. That we can see an earlier generation's social biases and scientific errors should be no cause for self-congratulation.

The last decade has witnessed a transformation in the way scholars think about the history of eugenics. But the new emphasis on the diversity of eugenicists' backgrounds and aims, as well as eugenics' close association with genetics, has barely touched what we see on TV or read in magazines. Popular discussions of the history of eugenics have come increasingly to diverge from scholarly accounts. This book aims to bridge the gap between expert and lay understandings of the history of eugenics and to investigate eugenics' links to human genetics. It is written in the hope that a more complex historical understanding will also enrich debate on contemporary choices.

Evolutionary Anxieties

N INETEENTH-CENTURY BRITAIN was a society racked by social turmoil and bitter class conflict. By the 1840s, the idea of "two nations"—one constituted by brutalized and menacing urban workers, the other by their cloistered and decadent rulers—had become a stock image. Victorian thinkers struggled to explain the apparent descent of so many city dwellers into lives of pauperism, violence, and crime. In Britain as in Continental Europe, these thinkers often attributed rampant disease and disorder to "degeneration," a vaguely defined process of biological and social decline (Pick 1989, 21).

Charles Darwin and Francis Galton were products of their time and class. Their work both reflected and reinforced the sense that civilization was under siege. Like most Victorian gentlemen, they were alarmed by the fecundity of the lower classes. Darwin, however, was considerably more cautious than Galton in advancing solutions. Whereas Galton was assertive, caustic, openly antireligious, and zealous, his cousin was generally tolerant, patient, and loath to offend. These temperamental differences were reflected in their responses to social developments. Galton wrote of the need for policies to control human breeding, Darwin of the need for more education in the principles of breeding. But if Darwin's views were mild in comparison with Galton's, his theories did much to fuel the fears that made his cousin's proposals compelling. In this chapter we will see how their work converged to intensify anxieties about degeneration and to focus those anxieties on the reproductive behavior of the urban poor.

Charles Darwin's Conflict in Attitudes

In the *Origin of Species* (1859), Charles Darwin had argued that evolution proceeded through the "survival of the fittest." But what is fitness? Darwin's own answer was ambiguous. Like biologists today, he sometimes

22

equated fitness with reproductive success. The fittest were simply those who left the most offspring. But Darwin more often employed the term colloquially; the fittest were the biggest, the strongest, the smartest, the swiftest. They were the physically and mentally "best," by a conventional standard of value.

If certain traits were always advantageous in the struggle for existence, evolution would be directional. And if the traits were prized, its direction would be up. In a global and not just a local sense, selection would constantly make organisms better and better. That is what Darwin thought, at least most of the time. "As natural selection works solely by and for the good of each being, all corporeal and mental endowments will tend to progress towards perfection," he wrote in the *Origin* (1859, 489).

At times, he also wrote in a different vein. On progress (as on some other matters) Darwin was conflicted. In the margins of his copy of an 1844 evolutionary treatise he noted, "Never speak of higher and lower in nature"—the maxim of most contemporary biologists. While he sometimes broke this vow, he never thought of progress as linear. Darwin's basic metaphor was a tree, not a ladder or chain. But if trees branch, they still grow upward. That is where Darwin, in general, thought evolution was headed. Contemporary evolutionists may insist that natural selection involves local adaptation to changing environments, a process with results devoid of moral meaning. They may even denounce the whole concept of progress as "noxious, culturally embedded, untestable, nonoperational [and] intractable" (Gould 1988, 319). But Darwin was a man of his time. Like other Victorians, he assumed that progress was natural, that all would work out for the good. Both his scientific and popular audiences certainly read him as a progressionist. They understood selection as a force tending to the constant improvement of plants and animals.

The human animal, however, seemed to be a special case. Darwin and his contemporaries worried that with respect to us, progress was not guaranteed. Acting on our moral sentiments, which were themselves the products of selection, we interfered with its beneficent culling of the weak. Civilized societies protected the physically and mentally vulnerable and thus prevented selection from doing its work.

The prospect of mental and moral decline was particularly alarming. More of the physically frail might survive, but they did not actually outbreed the strong. The weak in mind and character were another matter. The most eminent members of society seemed to produce the fewest offspring, the reckless and stupid the most. If this trend continued, the direction of evolution would be reversed. To Darwin's contemporaries, what to do about it—if anything—was a burning question.

Darwin's own attitudes were strongly influenced by his five-year voyage around the globe on the HMS *Beagle*. From 1831 to 1836, he served as naturalist and companion to the ship's captain, Robert Fitzroy. The voyage was notable for many reasons: it provided Darwin with numerous useful facts that he would later employ in the *Origin*, won him the respect of leading scientists, greatly boosted his self-confidence, reinforced his commitment to science, and undermined his Christian religious beliefs. His *Beagle* experiences also convinced him that the distance separating humans and animals was vastly less than his Victorian contemporaries imagined. By the time the voyage ended, Darwin no longer thought of humans as unique.

Of all his *Beagle* experiences, the one that most affected Darwin's view of human evolution was his encounter with the inhabitants of Tierra del Fuego, the cold and barren region at the southernmost tip of South America. Forty years later, he wrote of his astonishment on first seeing these people: "He who has seen a savage in his native land will not feel much shame, if forced to acknowledge that the blood of some more humble creature flows in his veins." Darwin remarked that he himself would prefer to be descended from a heroic monkey who saved the life of his keeper or a baboon who rescued his comrade from a pack of dogs than "from a savage who delights to torture his enemies, offers up bloody sacrifices, practices infanticide without remorse, treats his wives like slaves, knows no decency, and is haunted by the grossest superstitions" (1871, 2:404–5).

On a previous voyage, Captain Fitzroy had brought four Fuegians back to England in hopes of teaching them Christian religion, the English language, and the use of tools. If the experiment worked, the natives would return home as missionaries. Although one died of a smallpox vaccination, the remaining three—now named York Minster, Jemmy Button, and Fuegia Basket—seemed to adjust remarkably well. They learned some English (Fuegia also learned a little French) and gave every appearance of having converted to Christianity. During the summer, they were even received at court, Queen Adelaide presenting Fuegia Basket with her own ring and bonnet (Marks 1991, 43–45).

In December 1832, the Fuegians were returned home in the company of a British missionary. Darwin was impressed by his encounters with these Anglicized natives: "I was incessantly struck, whilst living with the Fuegians aboard the 'Beagle,' with the many little traits of character, shewing how similar their minds were to ours; and so it was with a full-blooded negro with whom I happened once to be intimate," he would recall many years later (1871, 1:232; see also 34). But as the *Beagle* approached shore, Darwin caught his first glimpse of the native inhabitants. He was horrified. He could not connect these barbarous natives with the passengers aboard their ship. In the diary he kept at the time, Darwin wrote that "3 years has been sufficient to change savages, into, as far as habits go, complete & voluntary Europeans" (Keynes 1988, 143). But these Fuegians seemed like wild animals with dirty and nearly naked bodies, long and tangled hair, painted bodies, talking so rapidly (in what sounded to Darwin like gibberish) that their mouths frothed, and gesturing wildly. Their property consisted only of bows, arrows, and spears and their diet of shellfish, along with seals and some birds. They had no gardens or real tools and only primitive canoes. The Fuegians were also fiercely aggressive, ready "to dash your brains out with a stone." In his diary, Darwin remarked: "I would not have believed how entire the difference between savage & civilized man is. It is greater than between a wild & domesticated animal, in as much as in man there is greater power of improvement" (Keynes 1988, 122–25, 129–30, 134–39).

Darwin ended the voyage convinced that Fuegians were closer to apes than to Englishmen. The experiment ultimately ended in disaster. As soon as the *Beagle* departed, the missionary's possessions were plundered. Even worse, York Minster and Fuegia Basket quickly shed their English clothes and manners. Darwin was also greatly affected by a shipmate's description of a child dashed against the rocks and left to die for dropping a basket of seagull eggs and by a report that the Fuegians were cannibals who killed and ate their old women in preference to dogs. He wrote in the notebooks he kept on his return to England: "Compare the Fuegians and the Ourang-outang, and dare to say differences so great" (Gruber and Barrett 1974, 296).

At the time of the *Beagle* voyage, Darwin was still a creationist; that is, he believed that God created plants and animals in their present form. Shortly after it ended in 1837, he had become a convinced materialist, committed to explanations of all phenomena—including human origins—in terms of natural law. In September 1838, Darwin read the English economist Thomas Malthus's *Essay on the Principle of Population*. His reading of Malthus, who argued that more organisms were born than could possibly survive, prompted the realization that the pressure of population on scarce resources would generate a fierce struggle for existence; given the intensity of the struggle, even minute advantages would increase the possessors' chances of surviving and reproducing their kind. Darwin called this process "natural selection" to emphasize the analogy with the familiar practice of conscious selection employed by domestic breeders.

As soon as Darwin grasped the concept of natural selection, he began

FIGURE 2.1   Woollya Cove, Jemmy Button's home. Engraving by T. Landseer, in Robert Fitzroy, vol. 2 of *Narrative of the Surveying Voyages of His Majesty's Ships Adventure and Beagle between the Years 1826 and 1836* (London, Henry Colburn, 1839), p. 208. Photo courtesy of Widener Library, Harvard University.

to reflect on its implications for humans. Thus, long before writing the *Origin*, his comprehensive materialism led him to accept the continuity of humans with other animals—in morals, emotions, and behavior as well as physique. But it was not until 1871 that Darwin was willing even to state that humans evolved, much less to discuss any social implications of natural selection. He did not want to intensify the controversy he knew the *Origin* was anyway certain to prompt. Thus the book included only a single, enigmatic reference to humans: "Light will be thrown on the origins of man and his history" (1859, 488).

The omission fooled almost no one. Both friends and enemies recognized that if Darwin's theory were true, it had implications for our understanding of "man's place in nature" (a phrase popularized by the naturalist Thomas Henry Huxley in 1863). Darwin was immediately assailed for reducing humans to brutes. The uproar reinforced his reluctance to speculate on the social meaning of natural selection. For the next ten years, he defended his theory while avoiding any reflection on his own species. His peers, however, were less inhibited. In the 1860s, they published a raft of works on the relevance of evolutionary theory to the human past and future.

Alfred Russel Wallace: Selection and Socialism

One of the first to explore this question was Alfred Russel Wallace, the cofounder of the principle of natural selection. In his 1864 article "The Origin of Human Races and the Antiquity of Man Deduced from the Theory of 'Natural Selection,'" Wallace attempted to resolve the long-standing dispute between monogenists (who argued that all humans are descended from a common ancestor, and thus constitute a single species) and polygenists (who maintained that human races have different origins, and thus belong to separate species [Durant 1979, 40]). Wallace proposed that human evolution had passed through an early stage of purely physical development, during which distinct races appeared, and a later stage, when selection acted mostly on mind. The monogenist and polygenist positions could thus be made consistent; although the races had a common origin, divergence had occurred in a distant past before the evolution of humans' most distinctively human characteristics: intelligence and character. Once selection began to act on the brain, humans became capable of transcending their physical environment, and the evolution of human physical form practically ceased.

Wallace argued that selection favors societies whose members are rational and altruistic. He reasoned that sympathy, cooperativeness, and foresight would provide an advantage in the struggle of one group with another; groups in which these traits predominated would prosper, while their rivals would fade and finally disappear. In his view, higher races would crowd out "the lower and more degraded" in a constant process of intellectual and moral improvement. As humans became ever more perfectly adapted to the natural and social worlds, government and restrictive laws would disappear, the advance of human moral faculties having rendered them superfluous. The earth would be converted "into as bright a paradise as ever haunted the dreams of seer or poet" (1864, 26). Thus Wallace deemphasized the struggle for existence within human societies, while stressing the impact of struggle among them.

Primitive peoples would also have vanished, selection having guaranteed the extinction of all "those low and mentally undeveloped populations with which Europeans come in contact." According to Wallace, "The red Indian in North America, and in Brasil; the Tasmanian, Australian and New Zealander in the southern hemisphere, die out, not from any one special cause, but from the inevitable effects of an unequal mental and physical struggle" (1864, 21). Darwin regarded the essay as the best ever to have appeared in the *Anthropological Review,* and he marked this particular passage (Greene 1921, 103–4).

Wallace himself was soon to have second thoughts. Unlike Darwin and most other British naturalists, he came from a very poor family and left school at age 14 to earn a living. His scientific research was mostly accomplished during two extended expeditions to the Amazon and the Malay archipelago, where he collected specimens, selling the duplicates to cover his costs. Wallace found much to admire in native societies, and the poverty and inequality of his own society appalled him. "The more I see of uncivilised people," he wrote in a typical passage, "the better I think of human nature on the whole, and the essential differences between so-called civilised and savage man seem to disappear" (quoted in Durant 1979, 43). Wallace's experiences of "native" and "civilized" societies were thus in extreme tension with his view that selection guaranteed the expansion of the latter at the expense of the former. Moreover, he found his article cited in support not only of colonialism but of a dog-eat-dog capitalism that he despised (Durant 1979, 44).

Thus, in an important essay, "On the Failure of 'Natural Selection' in the Case of Man," the Scots essayist William Greg agreed with Wallace that the process of natural selection operated in the struggle among tribes, nations, and races. "Everywhere," he wrote, "the savage tribes of mankind die out at the contact of the civilised ones." But Greg drew a conclusion very different from Wallace's about the role of selection in "civilized societies." Greg considered the struggle for survival beneficent and deplored the agencies that kept it in check. As a consequence of medicine and indiscriminate charity, he argued, the least valuable individuals and classes were now outbreeding the best. Among wild animals and savages, the sick and maimed succumbed, whereas in civilized Britain, they were cared for and allowed to propagate. "The indisputable effect of the state of social progress and culture we have reached," he asserted, "is to counteract and suspend the operation of that righteous and salutary law of 'natural selection' in virtue of which the best specimens of the race—the strongest, the finest, the worthiest—are those which survive, surmount . . . and propagate an ever improving and perfecting humanity." The pampered rich and idle poor both reproduced at will. Aristocrats were notoriously inbred and often mentally weak but, being rich, could breed as much as they pleased. Paupers and imbeciles, being too stupid and improvident to think of their future, did the same. Only the middle class—"the true strength and wealth and dignity of nations"—postponed or abstained from marriage (1868, 356, 360). In a sensible republic, only the elite would be allowed to reproduce. Alas, British society was headed in just the opposite direction. Darwin underlined and annotated Greg's article and marked it "keep" (Greene 1981, 109).

Wallace was an advocate of the social reforms that Greg and others condemned for weakening the race. At this time of increasing unease with the implications of his own account of human evolution, Wallace attended his first séance and was converted to spiritualism (the belief that it is possible to communicate with the dead, who survive as spirits). He now believed that "forces and influences not yet recognized by science" had guided the evolution of humans from animals. If evolution were steered by a higher power, it could progress toward its goal without need of a fierce competitive struggle (Durant 1979, 47).

In "The Limits of Selection as Applied to Man," Wallace spectacularly reversed his earlier position. He now argued that natural selection could not account for the development of certain human physical and mental traits. Primitive peoples had no use for a conscience, refined emotions, or abstract thought. Selection could not produce either the higher mental and moral qualities or hairlessness, since they had no value for survival in the wild. Yet, from the size of their brains and occasional evidence of conscience, it was obvious that primitive peoples possessed the same intellectual, emotional, and moral capacities as modern Englishmen. "Natural Selection could only have endowed savage man with a brain a little superior to that of an ape," asserted Wallace, "whereas he actually possesses one very little inferior to that of a philosopher" (1870, 356). He thus concluded that a higher power must have guided human evolution. Wallace's utilitarian objections to natural selection probably reflected his understanding that a frankly spiritualist case would have been ineffective. Having failed to persuade his peers to take spiritualism seriously, however, he was forced to exclude it from his discussion of human evolution (Kottler 1974). In any case, his reversal on human evolution came as a shock to Darwin, who voiced to Wallace the hope that "you have not murdered too completely your own and my child." In response, Darwin was finally prompted to publish theories that he had suppressed for over a decade. His two-volume *Descent of Man, and Selection in Relation to Sex* appeared finally in 1871.

Galton's Objections

That book was also profoundly influenced by Galton's work on the heritability of mental and moral traits. In a two-part article published in *Macmillan's Magazine* in 1865, Galton had bluntly acknowledged (as had Darwin in the *Origin*) that almost nothing was known of the laws of heredity. He argued, however, that we can recognize that like produces like without understanding the mechanics of inheritance. Both in humans and in brutes, it is clear that parents and offspring tend to resemble each other. And this resemblance applies as much to mental as to physical features.

To establish his point, Galton consulted reference works containing biographies of eminent men. He found that high achievement runs in families. Men distinguished in the sciences, arts, and public life were much more likely than the public at large to have fathers who were themselves eminent. Galton considered the possibility that social advantages explained these statistics. He understood that "when a parent has achieved great eminence, his son will be placed in a more favourable position for advancement, than if he had been the son of an ordinary person" (1865, 161). In the end, however, he dismissed the importance of this fact. Social inheritance might explain distinction in the legislature or army but not, he argued, in science, literature, or the law. Really capable people in these fields would overcome every hindrance to success. Developing this theme, Galton later argued that although culture and education were much more widely spread in America than in England, Americans produced few first-class works of literature, philosophy, or art. If English society were as egalitarian as the American, he concluded, it would "not become materially richer in 'highly eminent men'" (1869, 40).

However, many people were unconvinced by the claim that eminence reflected natural abilities. In response to criticism that he overstated the case for heredity, Galton published *English Men of Science: Their Nature and Nurture* (1874), a work based on data from questionnaires sent to 180 especially distinguished members of the Royal Society. Once again, the incidence of eminence among the relatives of these scientists was found to be much higher than among those with comparable educational backgrounds. The American Frederick Adams Woods established that in his country, too, prestige ran in families. On Woods's calculation, the 46 celebrities honored by tablets in the Hall of Fame "are from five hundred to a thousand times as much related to distinguished people" as was the ordinary American. Woods argued that since opportunities were freer in America, and social lines less strictly drawn, America provided an even better test than England of the inherited nature of exceptional ability. In his view, the results clearly vindicated Galton's thesis that success "is nearly all a matter of heredity" (1913, 448, 452). (Darwin advanced an analogous argument in respect to women. "The chief distinction in the intellectual powers of the two sexes," he wrote, "is shewn by man's attaining to a higher eminence in whatever he takes up, than woman" [1871, 2:327].)

The abilities cited by Galton included specific talents and traits of character and personality as well as general intellectual power. Indeed, everywhere that Galton looked he found the influence of heredity. Tuberculosis, longevity, madness, susceptibility to the poisonous effects of opium, gregariousness, an aversion to meat: all were inherited. "So is a craving for drink, or for gambling, strong sexual passion, a proclivity to pauperism, to crimes of violence, and to crimes of fraud" (Galton 1865, 320).

This argument did not find ready acceptance. A few scientific reviewers of Galton's 1869 book *Hereditary Genius* (an elaborated version of the articles) were enthusiastic. But reviewers in political, literary, and (especially) religious journals were less impressed. Even those who were intrigued by Galton's claims tended to be skeptical of his theory and methods (Forrest 1974, 101; Gökyiḡit 1994, 219). Galton's wife noted in her diary: "Frank's book 'Hereditary Genius' published in November, but not well received" (Cowan 1977, 139). Neither of his books on eugenics sold many copies, either in Britain or in America. The general indifference is partly explained by Galton's emphasis on the need for more intellectuals rather than fewer paupers and imbeciles; that is, on what he would later call "positive" as opposed to "negative" eugenics. The latter had much broader appeal, especially in America. But the lukewarm reception is also explained by its author's scorn for conventional theological, moral, and scientific views.

On the scientific side, Galton challenged basic assumptions about the nature of heredity. As we will see in the next chapter, most of his contemporaries assumed that traits of character and personality were innate. In the nineteenth century, however, "innate" did not imply "determined," for it was generally believed that the environment shaped heredity. Thus an inherited trait could be suppressed or redirected by changes in the condition of life. Galton insisted on a sharp disjunction between heredity and environment, and he thought the former all that mattered. "Will our children be born with more virtuous dispositions, if we ourselves have acquired virtuous habits?" he asked. "Or are we no more than passive transmitters of a nature we have received, and which we have no power to modify" (quoted in Cowan 1977, 142). The answer to the second question was yes.

Galton's views were at least as hard to reconcile with conventional religion. In his memoirs, Galton recalled that the *Origin* had dispelled his Christianity like a "nightmare" exposed to the light of day. "Its effect was to demolish a multitude of dogmatic barriers by a single stroke, and to arouse a spirit of rebellion against all ancient authorities whose positive and unauthenticated statements were contradicted by modern science" (1908, 298). By the 1860s, he was openly skeptical and anticlerical. In "Hereditary Talent and Character," Galton explained both religious sentiments and human reason as products of natural selection, denied the existence of the soul, and disparaged the doctrine of original sin. He insisted that flawed human nature is not a curse visited by God but a matter of biological inheritance. According to Galton, moralists assume that humans are born with a flawed nature. In the Christian view, human nature is fundamentally tainted with sin, which prevents people from doing the things that they know to be right. Human beings have noble intentions but are too weak to carry them through (1865, 327).

Galton proferred an alternative, naturalistic explanation for these base inclinations. Whereas Christianity explained imperfect nature as a consequence of our fall from grace, he saw it as a product of evolution. According to Galton, we are not fallen angels but recently risen apes. Our immediate ancestors were barbarians, fitted by natural selection to their conditions of life. We should not expect selection to have yet molded human nature to modern circumstances. As "animals suddenly transplanted among new conditions of climate and of food, our instincts fail us under the altered conditions." Civilization is recent; our nature has not yet caught up (1865, 327).

Galton's work also challenged the most familiar moral maxims, especially the claim that virtue will be rewarded with success. If achievement is wholly a function of (unequally distributed) natural abilities and traits of character, there is no apparent point to moral effort. In *Hereditary Genius*, Galton expressed his lack of "patience with the hypothesis occasionally expressed, and often implied, especially in tales written to teach children to be good—that babies are born pretty much alike and that the sole agencies in creating differences between boy and boy, and man and man are steady application and moral effort" (1869, 56). It seemed to follow that intellectual and moral struggles were fruitless. That was not a perspective easily accepted by Galton's contemporaries on either side of the Atlantic. As the reform-minded American sociologist Lester Frank Ward noted: "If nothing that the individual gains by the most heroic or the most assiduous effort can by any possibility be handed on to posterity, the incentive to effort is in great part removed. . . . If, as Mr. Galton puts it, nurture is nothing and nature everything, why not abandon nurture and leave the race wholly to nature?" (1891, 291). Galton did not flinch from this conclusion. But his contemporaries did.

Moreover, if Galton were right, it seemed there were no grounds for assigning personal responsibility. Individuals do not deserve blame for their vices or credit for their virtues since these are beyond their control. (What matters is that these traits are determined, not that the agent is heredity. Were character wholly determined by environment, the implications for personal responsibility would be the same.) As Frederick Woods noted: "Heredity, environment and free will may be called the three rival claimants in the causation of mental and moral traits" (1913, 445). Galton denied that environment and will had any real effect on human action.

Given the number of established truths that Galton contested—that humans have souls, that their nature is sinful and reason a gift from God, that they can change their character and are responsible for their actions, that hard work pays off—it is not surprising that his work initially met with skepticism. One exception was Alfred Russel Wallace, who reviewed *Hereditary Genius* quite favorably in the scientific journal *Nature*. Another was Charles Darwin, who wrote Wallace to say that he agreed with "every word" of his review (Marchant 1916, 206). Darwin also sent his cousin a congratulatory note: "I do not think I ever in all my life read anything more interesting and original" (Darwin and Secord 1903, 41).

As we have seen, Darwin was already a convinced materialist, prepared by both his own experience and Wallace's 1864 essay to believe that human mental and moral traits were products of selection. He was also already inclined to the view that the mentally and morally highest races would be advantaged in the struggle for existence and thus ultimately displace the lower. These opinions were reinforced by his cousin's work. (Galton's views on race are typified by the comment that "there exists a sentiment, for the most part quite unreasonable, against the gradual extinction of an inferior race" [1883, 200–201].) But Galton also added something new and quite important to this mix: the idea that mental and moral differences among individuals (and not only populations) were attributable to differences in heredity. Darwin wrote to Galton: "You have made a convert of an opponent in one sense, for I have always maintained that, excepting fools, men did not differ much in intellect, only in zeal and hard work" (Darwin and Secord 1903, 41). While Darwin continued to think that moral effort could matter, he increasingly stressed the power of heredity. He also worried about the consequences if Galton were right about the fecundity of the lower classes. Alfred Russel Wallace noted that Darwin expressed gloomy views about the future on the ground that natural selection no longer operated in human societies and that the least fit were outbreeding the most (Wallace 1890, 51). Indeed, Darwin came to fear that the direction of evolution would be reversed.

Thus *The Descent of Man* cautioned its readers to pay at least as much attention to the pedigrees of their prospective mates as to those of their horses and dogs. It also warned of the consequences of society's protecting the weak and improvident. Citing Galton, Darwin noted that "if the prudent avoid marriage, whilst the reckless marry, the inferior members will tend to supplant the better members of society." Notwithstanding checks on the reproductive capacities of the poor (whose unhealthy living conditions produced high death rates) and criminals (who were often in jail), Darwin feared that society's worst elements would outbreed its best. "If the various checks . . . do not prevent the reckless, the vicious, and the otherwise inferior members of society from increasing at a quicker rate than the better class of men," he warned, "the nation will retrograde, as has occurred too often in the history of the world. We must remember that progress is no invariable rule" (Darwin 1871, 177). The book ends with a caution about the practice of protecting the weak:

> Man, like every other animal, has no doubt advanced to his present high condition through a struggle for existence consequent on his rapid multiplication; and if he is to advance still higher he must remain subject to a severe struggle. . . . Hence our natural rate of increase, though leading to many and obvious evils, must not be greatly diminished by any means. There should be open competition for all men; and the most able should not be prevented by laws

or customs from succeeding best and rearing the largest number of offspring. (403)

Darwin did not propose any practical programs either to prevent the reproduction of those deemed inferior or encourage it among their superiors. All the suggestions for restoring the reproductive balance seemed to him either unrealistic or repugnant. While he expressed hope that the "weak in mind and body" might be convinced to refrain from breeding, he did not think it likely. Nor did he believe that Galton's proposals to prod the gifted to produce more children would meet with more success. Galton's ideal he thought "grand" but probably also "utopian."

Darwin found harsher measures unacceptable. Breeding by undesirables could be reduced either by active measures such as segregation or by withdrawing charity and allowing the weak to succumb. As a Whig, Darwin was generally skeptical of state intervention, and he was too humane to accept equably an end to charity—even in the interests of advancing the race. Moreover, he himself had argued that the sympathetic impulses were products of natural selection. Thus his feelings were extremely conflicted. Immediately after noting that "hardly anyone is so ignorant as to allow his worst [domestic] animals to breed," Darwin retreated from the implication that humans should cull their own stock. We would do harm to "the noblest part of our nature," he argued, were we to check our sympathetic instincts. Neglecting the weak and helpless would involve us in "a certain and great present evil" for only a probable benefit. "Hence we must bear without complaining the undoubtedly bad effects of the weak surviving and propagating their kind" (1871, 168–69). Darwin was left with only the faint hope that dissemination of the principles of inheritance would lead the ablest members of society to produce more children and the least competent to exercise restraint.

Later Darwinians, less charitable in spirit and less squeamish about state intervention, came to a different conclusion. The mathematician Karl Pearson (1857–1936) was Galton's most famous disciple. Director of the Galton Laboratory for National Eugenics at University College, London, where he was also head of the Department of Applied Statistics, Pearson considered himself a socialist. His views were closer to what a later generation would call national socialism. Pearson claimed that progress depended on a harsh Darwinian struggle among nations and races. It is simply a brutal fact, he argued, that natural selection "produces the good through suffering, an infinitely greater good which far outbalances the very obvious pain and evil." In the interests of civilization, the white man "should go and completely drive out the inferior race" (1900, 23). Pearson was equally unflinching when it came to his compatriots. Following Darwin, he recognized that natural selection had produced the growth of human sympathy. Moreover, their compassion for each other had aided the British in their struggle with lesser breeds. But Pearson thought compassion had developed to a point where it had unfortunately "cried Halt! to almost every form of racial purification." In his view, "One factor—absolutely needful for race survival—sympathy, has been developed in such an exaggerated form that we are in danger, by suspending selection, of lessening the effect of those other factors which automatically purge the state of the degenerates in mind and body" (1907, 24–25). Pearson's harsh views resonated with those of many of his contemporaries, including Darwin's son Leonard, who served as president of the British eugenics society from 1911 to 1928.

Darwin at most flirted with eugenics. But his work provided the context that made Galton's views on heredity compelling. The claim that social failure results from bad blood engendered such alarm in the late nineteenth century because of its link to Darwin's theory of natural selection. Eugenics was transformed from abstract idea to social movement when it became attached to widespread assumptions about evolutionary progress and decline.

Wallace's Dissent

Of course there were dissenters. Among them was Wallace, whose spiritualism and deepening commitment to socialism eventually inclined him in a different direction. Wallace came to believe that negative eugenics would provide a specious scientific rationale for class distinctions and would serve to postpone urgently needed social reforms. "Segregation of the unfit," he remarked in 1912, "is a mere excuse for establishing a medical tyranny. And we have enough of this kind of tyranny already . . . the world does not want the eugenist to set it straight. . . . Eugenics is simply the meddlesome interference of an arrogant scientific priestcraft" (1912, 177).

Wallace's socialist leanings had crystallized in 1889 as a result of reading Edward Bellamy's just-published utopian novel, *Looking Backward*. Bellamy's tale is set in Boston in the year 2000. The city has been transformed. Economic inequities have vanished, and with them most social problems. All industry has been nationalized, and workers organized as an "industrial army." In every sphere, competition has been replaced by cooperation. Bellamy's vision of a peaceful and united society, without strikes, disorder, or crime, exerted immense appeal on both sides of the Atlantic. "Nationalist Clubs," devoted to propagating his message, sprang up all over the United States. At the turn of the century, *Looking Backward* was the third most popular book in America (after *Uncle Tom's Cabin* and *Ben Hur*) and was translated into more than 20 languages. Wallace regarded the book as a decisive refutation of every "sneer, every objection, every argument [he] had ever read against socialism" (1905, 266–67). It also strongly influenced his attitude toward human evolution and eugenics.

Bellamy had been much impressed by Darwin's concept of "sexual selection," first discussed in the *Origin* and later elaborated in the *Descent*. Darwin proposed the existence of a mechanism that would explain an apparent anomaly in his theory. Natural selection preserves only variants that are advantageous to their bearers. But some traits, such as the huge antlers of the Irish elk or the long, gaudy tail of the peacock, seem to be useless, or even liabilities, in the struggle for existence. They weigh animals down or make them more visible to predators. Male birds, which sing beautifully and are often brightly colored and adorned with ruffs and other ornamental displays, represent a particularly striking problem. How can natural selection explain showy characteristics that place their bearers at risk?

Darwin thought it could not. Instead, he explained these characters as a result of sexual selection. The large antlers of the elk are advantageous in combat between males for possession of females. More important than male combat is female choice. Peacocks have long, gorgeous tails because peahens prefer them; male birds sing to charm the females. Attractive males are chosen as mates and pass their traits to the next generation. Such characters represent a cost in the struggle for life but an advantage in the all-important contest to leave the most progeny.

The concept of female choice (unlike male combat) was extremely controversial. One of its severest critics was Wallace. He argued that the characters Darwin considered trivial or even detrimental were actually of great importance in the struggle for existence. Birds, he argued, have no aesthetic sense. Males are brightly colored in comparison with females because the latter need to be concealed while hatching in open nests. All of Darwin's examples, he argued, were better explained as a consequence of ordinary natural selection.

But after reading Bellamy, Wallace adopted just the reverse position in respect to human evolution. *Looking Backward* apparently provided Wallace with more than a compelling defense of socialism; it also furnished a mechanism by which human progress could be secured (Fichman 1981). In Bellamy's egalitarian Boston, women would rationally choose their mates. "For the first time in history," explains one of the characters,

> the principle of sexual selection, with its tendency to preserve and transmit the better types of the race, and let the inferior types drop out, has unhindered operation. The necessities of poverty, the need of having a home, no longer tempt women to accept as the fathers of their children men whom they neither can love nor respect. Wealth and rank no longer divert attention from personal qualities. Gold no longer "gilds the straitened forehead of the fool." The gifts of person, mind, and disposition—beauty, wit, eloquence, kindness, generosity, geniality, courage—are sure of transmission to posterity (1888, 179).

Although he dismissed the importance of sexual selection in other species, Wallace seized on it as a source of progress and an alternative to negative eugenics in his own. With Bellamy, he argued that equalizing wealth would spontaneously bring about the biological improvement of the human race. Women would choose mates who excelled mentally, morally, and physically. Slowly but surely, selection "would tend to eliminate the lower and more degraded types of man, and thus continuously raise the standard of the race" (Wallace 1900, 1:517). That argument would find many echoes among social radicals. Thus the American suffragette Victoria Woodhull lamented that "sexual selection has very little scope in our conventional system" (Martin 1891, 20). So did George Bernard Shaw and a host of other socialists. In their perspective, economic security would reinstate Darwinian sexual selection as the proper mechanism of improved human breeding. Darwinism most often reinforced conventional views of women's abilities and appropriate roles. But radicals also deployed Darwinism for their own ends. They argued that the future of the species was endangered by the continued subjugation of women.

The Darwin-Wallace controversy illustrates the extent of shared assumptions among nineteenth-century evolutionists, as well as areas of dispute. Wallace and Bellamy, as much as Galton and Darwin, believed that mental and moral differences were inherited and that "the only method of advance for us, as for the lower animals, is in some form of natural selection" (Wallace 1905, 389). None doubted that the very poor were biologically unfit. Unlike Galton and Darwin, Wallace and Bellamy admired the working and lower middle classes. But their attitude toward what Marx termed the "lumpenproletariat," or what in the twentieth century was known as the "social residuum," was decidedly unsympathetic. They saw this stratum as bearers of bad heredity, not victims of social circumstance. In Wallace's view, eugenic legislation was unnecessary, since neither the very rich nor very poor bred as rapidly as the worthiest classes. Moreover, in a socialist society, untrammeled sexual selection would constantly improve the species. Thus he and Galton differed on both the seriousness and source of the threat. But all the participants in this debate agreed that group as well as individual differences were largely attributable to heredity.

Darwin's *Origin* and *Descent* provided a biological framework for understanding unsettling social problems that was broad enough to accommodate widely divergent approaches. Conservatives, socialists, and liberals all deduced social policy from the facts of evolutionary biology. But their views of the biological facts were strongly conditioned by their politics. Who were the "fittest"? There could be no objective answer. The socialist Wallace, a self-educated surveyor from an impoverished family, thought the qualities making for civic worth resided mostly in workers and shopkeepers. Galton and Darwin disagreed. They thought it resided in other professionals.

Few who wrote on the social implications of evolution (like those who write on any subject) came from backgrounds similar to Wallace's. The commentators were mostly professionals—doctors, lawyers, teachers, ministers, scientists—with university educations, from comfortable families. They tended to see the world as Galton and Darwin saw it. It was a world in which the fecundity of the "lower orders" was a source of great anxiety.

The sense of alarm was heightened by new conceptions of heredity. In 1883, the German cytologist August Weismann (1834–1914) argued that hereditary elements were impervious to changes in environment. This "hard" view of heredity was reinforced in 1900 with the rediscovery of the work of Gregor Mendel. The countries that adopted the harshest eugenic policies were those where Weismann and Mendel received their warmest reception. The next chapter explains why.

3

From Soft to Hard Heredity

IN THE LATTER half of the nineteenth century, eugenicists knew almost nothing about the physiological nature of heredity or the rules governing transmission of hereditary characteristics. Francis Galton was able to demonstrate a correlation between the traits of parents and offspring. But neither he nor his British followers (known as "biometricians") could precisely identify patterns of inheritance.

The physiological mechanisms of heredity were even more obscure. Until the 1880s, it was generally assumed that characteristics acquired by organisms during their lifetimes were transmissible to their progeny. This principle was given a behavioral twist by the French evolutionist Jean Baptiste de Lamarck (1744–1829). In his *Philosophie zoologique* (1809), Lamarck suggested that changing conditions create new needs, which animals alter their habits to satisfy. The resultant use and disuse of organs elicits physiological changes that over time become heritable. Plant (and some animal) structures are also directly modified through exposure to new environments.

Lamarck's "soft" view of heredity was embedded in a broader evolutionary theory that was soon discredited. But the principle of inheritance of acquired characteristics survived, as did Lamarck's notion that changes in heredity are linked to changes in habit. It followed from these assumptions that parents' reckless behavior, if chronic, would produce moral and physical weaknesses in their offspring. Parents were counseled by physicians and popular health manuals that anger, anxiety, bad temper, jealousy, and other negative emotions would produce constitutional ills in the child; thus conception should occur only when both parents were feeling tranquil and tender (Rosenberg 1976, 27). Since gestation and nursing were critical periods, mothers in particular needed to remain cheerful. (This advice reflected the common

40

assumption that females contribute more to temperament, males to intellect.) "The brain of the unborn child is powerfully influenced by the thoughts of the mother," asserted the author of *The Parent's Guide; or, Human Development through Inherited Tendencies.* Thus pregnant women should "cultivate generous feelings and noble aspirations" to their fullest extent. Moreover, the mother's image of the father during gestation strongly marks the character of her child; if she dwells on him "not only with ardent affection, but with proud admiration of his noble nature, then the offspring will be so many copies of the father— spiritualized and enlarged by the glorified imagination of the mother" (Pendleton 1871, 15, 20). Today "Lamarckism" is in disrepute, but during the nineteenth century such beliefs were uncontroversial.

Although Charles Darwin's theory of evolution by natural selection can be sharply distinguished from Lamarck's, Darwin's theory of heredity, which he called "pangenesis," follows the prevailing nineteenth-century view that acquired characters are heritable. Pangenesis accounts physiologically for how such inheritance might work: minute particles thrown off by various cells circulate through the body and ultimately concentrate in the "germ cells." This process explains how changes in parents' bodies could be manifested in their offspring.

Pangenesis had many critics, however. Among them was Francis Galton, who was also the first to reject wholesale the principle of inheritance of acquired characters. Galton claimed that the hereditary material was transmitted unaltered from one generation to the next. But he had no experimental evidence for his "hard" view of heredity. In 1883, the German cytologist August Weismann (1834–1914) provided such evidence. Weismann distinguished between two kinds of cells that constitute the human body: the germ cells, which are present in the gonads and give rise to sperm and eggs, and the "somatic cells," which are present in all other tissues of the body. On the basis of his research on the nature of the cell nucleus, he argued that the germ cells (or "germ plasm" as it was often called) are completely isolated from the somatic cells. While the somatic cells could be affected by the environment, the hereditary units of the germ plasm could not. They would be transmitted unaltered from one gene⸱ ⸱ ⸱⸱t No matter how much the parents worked to ⸱ minds, their heredity would be unchange⸱ the stronger or smarter.

Weismann's doctrine of the "continuity ⸱ many adherents in Britain and the United Sta⸱ (The French, in contrast, remained almost universally ⸱⸱⸱⸱⸱⸱⸱⸱⸱⸱⸱⸱ to Lamarckism.) Charles Davenport, a distinguished American geneticist and leading advocate of eugenics, explained that "if we study the pedigrees of [asocial] men carefully (and many of them have been studied for six or seven generations) we trace back a continuous trail of the defects until the conclusion is forced upon us that the defects of this germ plasm have surely come all the way down from man's ape-like ancestors, through 200 generations or more" (1912, 89). The trial lawyer Clarence Darrow made the same point more colloquially in the notorious Leopold and Loeb case of 1924, which involved the thrill-killing of a young boy. He told the jury that his client's genes were at fault. "I do not know what remote ancestor may have sent down the seed that corrupted him," Darrow exclaimed, "and I do not know through how many ancestors it may have passed until it reached Dicky Loeb. All I know is, it is true, and there is not a biologist in the world who will not say I am right" (quoted in Bryan and Bryan 1925, 546). (When not defending clients, Darrow ridiculed the "utter absurdity of tracing out any given germ-plasm or part thereof for nine generations, or five, or three." Indeed, Darrow was one of the few non-Catholic critics of eugenics in the 1920s [Darrow 1925, 153; see also Darrow 1926].)

The followers of Lamarck and Weismann were equally hereditarian; each attributed pauperism, licentiousness, criminality, and other deviant behaviors to defective heredity. But Lamarckians believed that bad environment was responsible for bad heredity and were therefore prone to support social reforms. Indeed, Lamarckism seemed to strengthen the case for correcting unhealthy conditions and habits. On the negative side, it was assumed that if these were not changed, symptoms would become progressively more severe in succeeding generations. On the positive, improvements in education, housing, and public health would enhance not just the current but future generations (Haller 1985, 24). The laws of heredity, wrote one Lamarckian, are "a proclamation to every man to institute a reform" (Rosenberg 1976, 36; see also Desmond 1989, 4).

That reformist spirit pervaded the first generation of American "family studies." Beginning in the 1870s, researchers identified rural clans beset by such social ills as pauperism, prostitution, insanity, and crime. From public records and from interviews with inmates of state institutions, their family physicians, relatives, neighbors, employers, teachers, and friends, investigators constructed pedigrees demonstrating the recurrence of undesirable traits in several generations of the same family. If a trait ran in families, it was assumed to be inherited.

The family-study method had been pioneered by Galton, who used it to demonstrate that *high* achievement runs in families. The American research had a different slant. It was generally based on field investigations rather than information culled from reference books, and it most often traced the lineage of social failure rather than success. Some Americans were interested in the genealogies of musicians, mathematicians, naval officers, and other distinguished persons. But more were concerned with the prostitutes, paupers, drunks, feebleminded, and insane who were cared for in state institutions or who supplemented casual work with petty crime and public charity. This "social residuum" represented an insupportable charge on the public purse. The family-study researchers aimed to understand the hereditary character of social pathology in the interests of limiting the cost of welfare (or what was then called "outdoor relief") and custodial institutions.

The first and best-known study was conducted by Richard L. Dugdale, a member of the executive committee of the Prison Association of New York, who had been appointed a one-person committee to investigate the state jails. The published version of his 1874 report to the New York legislature, *"The Jukes": A Study in Crime, Pauperism, Disease, and Heredity*, had immense and lasting impact. Dugdale began by interviewing prisoners in 13 county jails. In the course of his research, he found a family marked by criminal careers that could be traced back through several generations. The males in what Dugdale termed the "Juke" family had been convicted of assault, murder, rape, burglary, and cruelty to animals, among other offenses. Their local reputation was fearsome. Dugdale wrote, "They had lived in the same locality [in upstate New York] for generations, and were so despised by the reputable community that their family name *had come to be used generically as a term of reproach*" (1877, 8).

Dugdale's description was horrifying, but his message was hopeful. He presumed that the Juke family had inherited a proclivity for criminal behavior, but he also believed that a hereditary tendency to crime could be easily deflected. Dugdale's optimism reflected assumptions about the nature of both criminality and heredity. He was a convinced Lamarckian, who believed that "environment tends to produce habits which may become hereditary... if it should be sufficiently constant to produce modification of cerebral tissue" (66). Thus parents may transmit an inclination to pauperism, promiscuity, or theft, among other social pathologies. But heredity is not destiny. Primitive tendencies will be either reinforced or deflected by education and experience. Changes in environment (especially when they occur in youth and are permanently maintained) can reverse inherited inclinations. That is why Dugdale could insist that "public health and infant education... are the two legs upon which the general morality of the future must travel" (119).

Criminal tendencies, he thought, are among the most easily redirected because, unlike harlots or paupers, criminals are vigorous and enterprising. Criminals' qualities would make for success in honest pursuits; their energies only need to be channeled in productive directions. But the prison system, which "masses an army of moral cripples," reinforces rather than deflects the hereditary tendency to crime (50, 113). Some criminals may be hopelessly recalcitrant, and these must be prevented from breeding. But most can and should be saved.

Dugdale's reformist conclusion is echoed by Frank Blackmar in his study of the "Smoky Pilgrims," a pauper and criminal family in rural Kansas. Its members lived in filthy hovels and supported themselves mostly by stealing, begging, and prostitution. Blackmar wrote that "they seem scarcely worth saving. But from social considerations, it is necessary to save such people, that society may be perpetuated. The principle of social evolution is to make the strong stronger ... [but] the weak must be cared for or they will eventually destroy or counteract the efforts of the strong. We need social sanitation, which is the ultimate aim of the study of social pathology" (1897, 65).

As Lamarckism lost ground, the family studies changed character. The later investigations adopted a much harsher tone; their recommendations tended to segregation and sterilization rather than kindergartens and improvements in sanitation and housing. Nineteenth-century theorists generally assumed that degeneracy was accompanied by high rates of infant mortality and sterility and was therefore self-limiting (Gelb 1990, 245). As we will see in the next chapter, their successors tended to assume the opposite: that such families were especially prolific. From this, combined with the assumption that heredity is immutable, it followed that the only effective solution to social problems was breeding from the fitter stocks. That view was expressed by the geneticist Charles Davenport: "Apart from migration, there is only one way to get socially desirable traits into our social life, and that is by reproduction; there is only one to get them out, by preventing their reproduction through breeding" (Davenport and Laughlin 1915, 4). A similar view was expressed by the American sociologist Lester Frank Ward, who concluded that "if they [Weismann and his followers] are right, education has no value for the future of mankind" (1891, 252). But Ward was a reform-oriented Lamarckian who hoped they were wrong.

Not every Lamarckian was democratic and egalitarian, nor every convert to Weismannism a reactionary. Indeed, Herbert Spencer, the leading proponent of Lamarckism in Britain, favored unrestrained capitalism, whereas Alfred Russel Wallace, its leading opponent, rejected capitalism in favor of a planned economy. Scientific theories are socially plastic;

they can be and frequently are turned to contradictory purposes. Thus we should not expect absolute correlations between scientific theories and social views.

Some Lamarckians championed segregation or sterilization on the grounds that the process of deterioration had gone so far as to be irreparable or at least too costly to reverse; thus individuals with bad heredity should be prevented from propagating their kind. The American suffragette Victoria Clafin Woodhull (1838–1927) was a Lamarckian, but she asserted in 1891 that "the best minds of today have accepted the fact that if superior people are desired, they must be bred; and if imbeciles, criminals, paupers, and [the] otherwise unfit are undesirable citizens, they must not be bred" (Martin 1891, 38). Likewise, Weismannians on the political left generally rejected the notion that bad heredity explained social ills. They thought heredity was fixed but did not view it as the cause of poverty or crime. Indeed, some Weismannians tried to turn the Lamarckian argument on its head. Thus the socialist Wallace argued that the case against inheritance of acquired characters was cause for relief, since it implied that neither the debauchery of the wealthy nor the sordid habits of oppressed workers need produce any permanent degradation (1900, 2:505).

However, few radicals were convinced by this logic. Thus, in Russia, the Bolshevik Revolution was followed by an upsurge of Lamarckian sentiment. To newly revolutionized students and workers, everything seemed possible to those with the will to make it so—an attitude that seemed much more compatible with Lamarck's views than with Weismann's. By 1948 (primarily for reasons related to its promise of quick results in agriculture), the inheritance of acquired characters had become official state dogma. Weismann's doctrine of the continuity of the germ plasm, which by then was generally accepted by geneticists, was condemned as having racist and reactionary implications. But Marxist scientists who accepted Weismannism denied the charge that it would give comfort to enemies of the proletariat. A. S. Serebrovsky in the Soviet Union, H. J. Muller in the United States, and J. B. S. Haldane, a member of the Executive Committee of the Communist Party in Britain, followed Wallace and advanced the counterclaim: not Weismannism but Lamarckism was reactionary, since it implied that working-class and colonial peoples, having lived in impoverished conditions, would now be genetically inferior to more powerful classes and nations. As Haldane wrote in response to one right-wing Lamarckian: "Reactionary biologists . . . who think that the unemployed should be sterilized, naturally use the theory of the transmission of acquired habits for political ends. It is silly, they say, to expect the children of manual workers to take up book-learning, or those of long-oppressed races to govern themselves. Laboratory experiments agree with social experience in proving that this theory is false" (1939, 243).

Haldane's argument was ultimately ineffective. It appealed to left-leaning scientists but not to their audience. The letters that appeared in the British Communist Party newspaper and in Marxist journals indicate that the party rank-and-file preferred the view that acquired characters were heritable. If certain groups suffered genetically as a result of living in deprived environments, that situation was remediable. The possibility of conscious intervention to improve heredity appealed to other political progressives and much of the public as well. Lamarckism resonated with deeply held attitudes about the value of efforts at self-improvement. Parents' strivings would be rewarded. Societies could pull themselves up biologically by their own bootstraps. Notwithstanding some exceptions among scientists, Lamarckism generally found favor with social reformers, who saw the alternative as a prescription for despair.

After 1900, the view that heredity was stable and fixed was buttressed by the view that it was governed by discrete "factors" (later termed "genes"). These factors maintained their integrity and did not become altered by blending. Working with edible garden peas, Mendel had demonstrated in 1865 that the hereditary material is transmitted intact from parent to offspring. Mendel also had established that some hereditary factors are expressed only when the same factor is inherited from both parents, while others are expressed even when only one factor is present. He designated the former factors "recessive" and the latter "dominant." Thus a pea plant carrying two copies of the dominant factor for yellow seeds (one received from each parent) would be indistinguishable in outward appearance from a plant carrying one copy of the factor for yellow and one for green seed. That a hybrid plant carried one recessive green factor was obvious when these plants were self-fertilized: the recessive trait reappeared in their progeny, in a ratio of three plants with yellow seeds to every one with green—the now famous 3:1 ratio.

For reasons that are still disputed by historians, Mendel's own contemporaries were unimpressed (Hartl and Orel 1992). When his work was rediscovered in 1900, the reaction was mixed (Hartl and Orel 1992). Some Darwinians concluded that Mendel's laws applied only to unimportant traits. Galton himself was one of the skeptics. His followers, the "biometricians," were also dismissive. They considered Mendel's laws true—as far as they went—but applicable only to uninteresting traits like seed color in garden peas. Unlike the Mendelians (who even-

FIGURE 3.1 "Mendel's Theatre," showing inheritance of hair color. Courtesy of the American Philosophical Society, Philadelphia.

tually won the dispute), the biometricians doubted their relevance to humans. Thus it is obvious that sympathy for eugenics did not depend on an acceptance of Mendelian genetics.

But Mendelism did have important effects. Mendel's internal factors followed laws that were entirely independent of the external environment. Mendelian analysis thus reinforced the contrast between the unchanging hereditary material and malleable body cells. For this reason, it seemed to strengthen Galton's and Weismann's arguments against Lamarck and conversely to weaken the view that biological inheritance was only a matter of influence and hence capable of being resisted or diverted. It represented another strand in the case for "hard" heredity.

It also appeared to establish that traits making for social success and failure were heritable. Francis Galton had been the first to demonstrate that mental traits ran in families. But he was unable to prove that his findings were explained by heredity. "All I can show," he wrote, "is that talent and peculiarities of character are found in the children, when they have existed in either of the parents, to an extent beyond all question greater than in the children of ordinary persons." Galton's facts were not disputed, but his explanation was. Many of his contemporaries thought social inheritance explained why "out of every hundred sons of men distinguished in the open professions, no less than eight are found to have rivalled their fathers in eminence" (Galton 1865, 158, 318).

Mendelian analysis seemed to seal the case for heredity. Mental and moral traits not only ran in families; they apparently followed a clear-cut Mendelian pattern. A plethora of American family studies, whose results were publicized in exhibits at state fairs and expositions, high school and college textbooks, and popular books and magazines, demonstrated the exact ratios expected on the assumption that mental defect resulted from a single, recessive gene. What more convincing proof of the hereditary transmission of traits than the oft-asserted "fact" that crosses between two normal parents always produced normal children, and crosses between two feebleminded parents always produced abnormal children, while the family lines resulting from matings between normal and feebleminded adults produced, on average, one abnormal for every three unaffected children (two of whom were assumed to be carriers of the defective gene)?

As we will see in the next chapter, if social worth was a reflection of an immutable heredity, the nation was clearly in trouble. A Harvard class barely reproduced itself, while paupers and petty criminals seemed to breed like rabbits. To "breed from the bottom" could end only in disaster (Popenoe and Johnson 1918, 139). The worst threat was unchecked feeblemindedness, for mental defect was considered the root cause of most social ills, such as pauperism, licentiousness, and crime. For a Lamarckian like Richard Dugdale, the solution to these ills lay in kindergartens and improvements in public health. Paupers, prostitutes, and criminals might suffer from bad heredity, but their heredity would improve with improved social conditions. His twentieth-century successors, who were mostly followers of Weismann and Mendel, had little faith in the effectiveness of social reform. Forty years after Dugdale's report *The Jukes* first appeared, Arthur Estabrook, a field researcher and collaborator of Charles Davenport at the Carnegie Institution, updated and reanalyzed the Juke family data. The message of Estabrook's *The Jukes in 1915* was the exact opposite of that drawn by Dugdale. But Estabrook's version is the one that carried the day. After 1915, the Jukes came to symbolize the futility of social change and the need for eugenic segregation and sterilization.

PRESENT-DAY DESCENDANTS OF ADA JUKE

Most of our social problems are produced by comparatively few families; the elimination of a few defective lines would result in an enormous reduction in our public bills.

FIGURE 3.2   "Present-Day Descendants of Ada Juke." From pamphlet by Marion S. Norton, "Selective Sterilization in Primer Form" (Princeton: Sterilization League of New Jersey, 1937); supplied by Garland E. Allen.

The Menace of the Moron

H ENRY H. GODDARD'S *The Kallikak Family: A Study in the Heredity of Feeble-Mindedness* (1912) chronicles the history of two family lines—one wholesome, the other a "race of defective degenerates"—both descended from the same male ancestor. Readers were assured that the family was real, though the Kallikak name was fictitious. To underscore the gulf between the family's depraved and respectable sides, Goddard had invented an alias by joining the Greek words for "beautiful" (*kalos*) and "bad" (*kakos*).

The Kallikak Family ultimately went through twelve editions (the last in 1939) and was nearly turned into a Broadway play (Smith 1985, 51). In the 1920s and 1930s, references to the Kallikaks are everywhere: in scholarly books and journals, college and even high school textbooks, serious and popular magazines. Richard Dugdale's Jukes and Goddard's Kallikaks jointly came to exemplify what a leading biology text labeled "Parasitism and its Cost to Society." After summarizing the results of both studies, the author asserted: "The evidence and the moral speak for themselves." Nevertheless, he went on to explain that hundreds of families like the Jukes and the Kallikaks continue to

> spread disease, immorality, and crime to all parts of the country. The cost to society of such families is very severe. Just as certain animals or plants become parasitic on other plants or animals, these families have become parasitic on society. They not only do harm to others by corrupting, stealing, or spreading disease, but they are actually protected and cared for by the state out of public money. Largely for them the poorhouse and the asylum exist. They take from society, but they give nothing in return. They are true parasites. (Hunter 1914, 263)

Since we cannot kill them off, as we would lower animals, we must prevent them from breeding. The studies were cited in other countries as well. In Denmark, which adopted the first European steriliza-

50

tion law, Goddard was one of the most frequently cited authorities on the inheritance of mental defect, and the Kallikak family figured prominently in eugenic literature (Hansen, in press). In Germany (where Goddard's book appeared in translation in 1914), the Kallikaks were equally famous (Proctor 1988, 99–100).

Today the Jukes and the Kallikaks are still often paired, though used to teach a markedly different lesson. The studies now mostly serve as examples of the depths to which science can sink; they are invoked to warn against the corruption of scientific judgment by social bias. But the coupling of Goddard and Dugdale was always misleading. In reality, they expressed sharply divergent scientific and social opinions. Those differences reflect a more general shift in attitudes about how to explain and to cure degeneracy. *The Kallikak Family* exemplifies the family studies that emerged in the wake of Weismann and Mendel. The genesis of Goddard's book reveals how social and scientific events converged to produce a terrible new threat—the prolific but largely invisible "moron" (a term coined by Goddard in 1910 from the Greek word for "dull" or "foolish")—that could be countered only by programs to control reproduction.

In 1906, Goddard was hired to open a psychological laboratory at the Training School for Backward and Feeble-Minded Children in Vineland, New Jersey. He found the existing data on mental defect of little value for understanding its cause. Family histories, when they could be obtained at all, were often biased by parents' ignorance or feelings of shame and thus could not provide a reliable basis for the construction of pedigrees (Zenderland 1992; Zenderland, forthcoming). Goddard concluded that field workers, trained to locate family members and diagnose their physical, mental, and moral condition, would do the best job of collecting data. He thought that "the people who are best at this work, and who I believe should do this work are women" (quoted in Gould 1981, 165).

He thus hired a former schoolteacher, Elizabeth Kite, to trace the family of a resident of the Vineland school. (The research was funded by wealthy philanthropist Samuel S. Fels, president of the company that manufactured Naphtha soap [Smith 1985, 45].) Deborah Kallikak had been admitted to Vineland at the age of eight and classified as a high-grade feebleminded or moron. According to the admission report, she washed and dressed herself, knew a few letters, understood commands, and could sew. Her behavior was more noticeably deficient; she was characterized as "careless in dress," "obstinate and destructive," and "not very obedient."

However, Deborah's conduct and skills both improved over time.

She learned to read (though with difficulty), add, play the cornet, and even sight-read music. By the time Goddard wrote his book, Deborah was a charming and beautiful woman, an excellent gardener, seamstress, and woodcarver, with a cheerful disposition and no obvious defects. Indeed, Goddard remarked on her teachers' unwillingness to admit even to themselves that she was feebleminded. Deborah would certainly not have appeared so to the casual observer. But as we will see, the morons' concealed character was the source of their threat.

Through interviews with Deborah's living relatives, Kite traced Deborah's genealogy back to a great-great-grandfather, Martin Kallikak. She found this family line appalling, with infant mortality high and drunkenness, pauperism, and sexual immorality rampant. The number of mental defectives was especially striking. On the basis of Kite's reports, Goddard conclusively judged 143 of the 480 descendants as feebleminded and 46 as normal, the rest being either unknown or doubtful.

In the course of her field work, Kite located another family of the same name but markedly contrasting character. These Kallikaks were all respectable members of their communities. The riddle of the two families was solved when Kite learned that during the American Revolution a young Martin Kallikak, then serving in the militia, had dallied with a feebleminded girl he met at a tavern. From their liaison issued an illegitimate son, Martin Kallikak Jr. (otherwise known as the "Old Horror")—the progenitor of Deborah's line. Martin Sr., who came from a good family, later married a respectable woman; from this union there ultimately issued 496 upstanding offspring. These Kallikaks had married into some of the state's most distinguished families, including descendants of colonial governors and signers of the Declaration of Independence. To Goddard, the family seemed "a natural experiment in heredity" (1912, 116). From that experiment, he concluded that feeblemindedness was hereditary ("the human family shows varying stocks or strains that are as marked and that breed as true as anything in plant or animal life") and the cause of virtually all social ills, including pauperism, criminality, prostitution, and drunkenness (1912, 12). Fortunately, an "ideal" solution was at hand: segregation of the feebleminded in institutions such as the one where he worked.

In his preface to *The Kallikak Family*, Goddard remarked that "some readers may question how it has been possible to get such definite data in regard to people who lived so long ago." After all, many of those included in Kite's pedigree were dead, some for as long as four or five generations, whereas others could not be located. But Goddard dismissed the problem on the grounds that "after some experience, the field worker becomes expert in inferring the condition of those

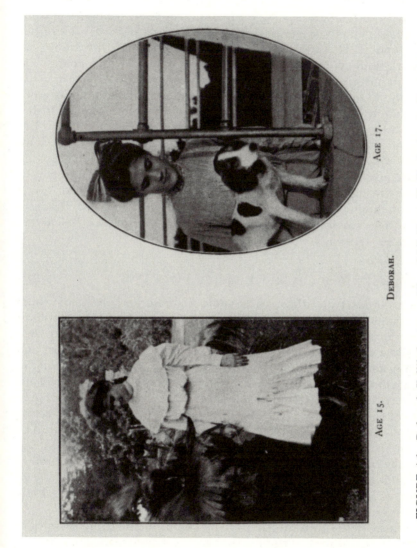

FIGURE 4.1 Deborah Kallikak at ages 15 and 17. From H. H. Goddard, *The Kallikak Family: A Study in the Heredity of Feeble-Mindedness* (New York: Macmillan, 1912).

persons who are not seen, from the similarity of the language used in describing them to that used in describing persons whom she has seen" (1912, 15).

Goddard did not acknowledge any problems in assessing the persons Kite did see. One chapter in the Kallikak book consists of a few of her actual reports. These were included both to make the story more entertaining and to "show something of her method, and enable the reader to judge of the reliability of the data" (1912, 71). That they do. From the standpoint of a modern reader, Kite's judgments were astonishingly swift, superficial, and subjective. Thus Kite comments that a 12-year-old girl should have been at school, "but when one saw her face, one realized it made no difference. She was pretty . . . but there was no mind there." The three children were said to have "the unmistakable look of the feeble-minded." The father of another family, "though strong and vigorous, showed by his face that he had only a child's mentality." For one boy "a glance sufficed to establish his mentality, which was low" (1912, 72–73, 77–78). *The Kallikak Family* data were clearly spurious.

Women and Eugenic Field Work

Kite's methods typify those of eugenic field workers, most of whom were trained by Charles Davenport and Harry H. Laughlin. Goddard collaborated with them on a manual for collecting and analyzing field data (Davenport et al. 1911). Between 1910 and 1924, Davenport and Laughlin taught basic principles of genetics to about 250 students in the Eugenics Record Office courses at Cold Spring Harbor. The students also learned how to conduct field studies and construct a pedigree chart (Allen 1986). More than 90 percent were female, in part reflecting Davenport's relatively positive attitude toward women and work: he favored increasing their professional opportunities (and thus was unusually ready to hire female assistants, as long as it did not interfere with childbearing ([Rafter 1988, 21]). So did Karl Pearson, who directed the two English laboratories devoted to eugenic research (Love 1979, 145–46).

In return, those who directed the research obtained the services of educated women who were prepared to do skilled academic work for extremely low pay. Many such women were available. In both Britain and America, universities had newly opened to females. In Britain, Bedford College for Women was founded in 1849; London University first allowed women to sit for its examinations in 1878 (Love 1979, 147–48). In the United States, Vassar College opened in 1865, Wellesley and Smith in 1875, and Bryn Mawr in 1893. In the 1880s, Harvard, Columbia, and Brown all founded annexes or sister institutions for women. At the same time, many other institutions became coeducational. By the 1890s, two-thirds of American colleges and universities admitted women, who constituted 36 percent of undergraduate students and 13 percent of graduate students (Muncy 1991, 4).

Universities were much readier to educate than to hire women. (It was expected that higher education would enhance their personal lives and roles as mothers, not qualify women to enter the work force.) In science, which was considered a "masculine" endeavor, the barriers to employment were especially high. Women who did find jobs were (as in all fields) shunted to specifically designated, subordinate positions that were low in both pay and status (Rossiter 1982, xvii). Sex segregation was justified in part on the grounds that women had special skills— cooperativeness, emotional sensitivity, perserverance, patience, close attention to detail—that suited them for particular kinds of work (Rossiter 1982, 52).

Women were considered especially well suited for eugenic field work. Eugenics' core issues were assumed by members of both sexes to fall within women's special domain. Women were also thought to possess exactly the right mix of social and analytic skills for family-study research. To obtain cooperation from family members, it was necessary to develop sympathetic relationships. Discharged patients and their relatives and friends would talk freely only if put at their ease (Davenport et al. 1911, 2). Women were thought particularly adept at building confidence. "Dropping in on a hot day and asking for a glass of milk or water, at once rouses friendly interest," explained Kite, who had taught science in public and private schools before she came to Vineland (quoted in Rafter 1988, 21). Women's ostensibly good intuition and sharp eye for detail were also essential if an individual's physical, mental, and temperamental traits were to be both swiftly and accurately assessed. However, women could only be employed by the Eugenics Record Office for three years. Since they were vigorous and intelligent, their employers considered reproduction their most important social duty (Barkan 1992, 72).

Although eugenic research represented one of the few entrees into science for educated women, the large number of female investigators cannot be explained simply by a lack of other career opportunities. Women were also drawn to eugenics by sympathy with its ideals, which explains why so many were involved in unpaid activities. In Britain, women constituted half the membership and a quarter of the officers of the Eugenics Education Society (Soloway 1990, 128). While women

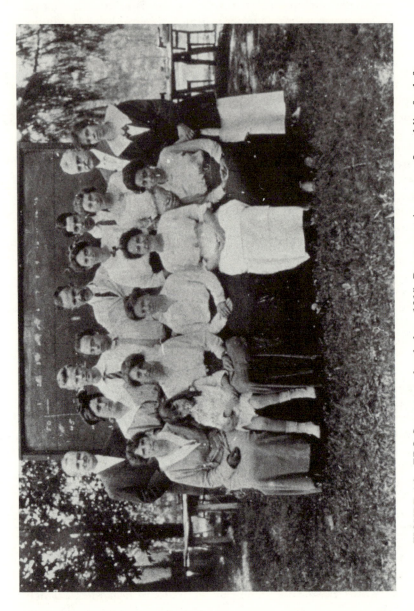

FIGURE 4.2 ERO Summer school class, 1915, Davenport in center, Laughlin far left. Courtesy of the H. H. Laughlin Papers, Pickler Memorial Library, Northeast Missouri State University.

in the United States played only a minor role in the national eugenics society, they were prominent in local groups (Kevles 1985, 64; Larson 1995, 71–79). Women were also very active in the Canadian eugenics movement (McLaren 1990). By the turn of the century, educated women had become extremely active in grassroots reform movements. Through participation in women's clubs, civic associations, and municipal and state politics, they worked to protect women and to shield the family from the rigors of the marketplace. Women organized to curb tenement homework, to end child labor and factory abuses, to pass laws providing for workmen's compensation, mothers' pensions, and child support, to trace deserting husbands, and to assist mothers so that they could stay at home with their children (Boris 1991). These efforts all involved a radical expansion of the state in the regulation of family life.

To many activist middle-class women, eugenics seemed a natural part of this wider movement to engage the state in new kinds of social reform. Eugenics' focus on the family and its theme of sacrifice on behalf of large impersonal ends especially resonated with women. The spirit of self-sacrifice was reflected in Elizabeth Kite's rejection of Goddard's offer to raise her salary. She said: "I fixed the rate at which I could work at the lowest possible figure at which I could leave home and I should despise myself forever if I accepted an increase of seventy-five cents a day" (quoted in Smith 1985, 51).

Rating Families

The summer school graduates were sent for a year's field training, mostly at institutions for the feebleminded, delinquent, or insane. They were armed with the *Trait Book*, which Davenport prepared in 1912 with assistance from the psychologists Edward L. Thorndike and Robert M. Yerkes. It listed hundreds of physical, occupational, and mental characteristics. These ranged from cataracts, cleft palate, curvature of the spine, color blindness, constipation, and consumption to "constancy in the sex realm," coolness in emergencies, cowardice, curiosity, cheerfulness, civility, and callousness—to choose just a few of the Cs. The data were catalogued at the Eugenics Record Office.

Field workers who participated in family studies prepared an "Individual Analysis Card" for every member of a family tree (whether living or dead). The worker was asked to evaluate the degree to which subjects were given to daydreaming or sticking to the facts at hand, held broad or narrow views, could profit from experience, possessed common sense, moral courage, and patriotism, could stand criticism, and could take a joke, among other characteristics (Davenport and Laughlin 1915).

The content of these cards reflected Davenport's assumption that most skills and aspects of temperament and character were independently inherited. Like Galton, he considered the fact that a trait ran in families proof of its heritability. Shiftlessness, "narcotism" (a taste for alcohol and drugs), extreme eroticism, artistic and literary ability, and love of the sea were only a few of the traits Davenport thought heritable. In his 1915 report *Nomadism, or the Wandering Impulse, with Special Reference to Heredity*, Davenport concluded that nomadism was probably a sex-linked recessive trait. Goddard's work convinced him that low mentality was a recessive trait as well; if a factor for normality were lacking in both parents, it would be lacking in all their offspring. It also followed that all the children of two feebleminded parents would be feebleminded themselves. "In view of this certainty," Davenport asserted, it would be a crime to let two affected persons marry (1911, 67).

However, it would be fine if they married normal people. Thus Davenport wrote that he could "well imagine the marrying of a well-to-do, mentally strong man and a high-grade feeble-minded woman with beauty and social graces which should not only be productive of perfect domestic happiness but also of a large family of normal happy children." "Cretins," as well as epileptics, the insane, hereditary criminals, and prostitutes, should be segregated during their reproductive years (1911, 257–59). But those of more normal mentality needed only to be educated in the facts of heredity. If they were potential carriers (as determined by family pedigrees), they should marry people who were not. Davenport's motto, "Strength should marry weakness and weak-ness strength," appalled some other geneticists, because the result would be an increase in the proportion of harmful genes in future generations.

Davenport attributed degenerate clans to the clustering of various bad traits, of which feeblemindedness was only one. He wished to separate the problems of mental defect and emotional control. Goddard, however, aimed to join them. He argued for the inheritance of a general mental ability and attributed most crime, sexual licentiousness, and idleness to its absence. *The Kallikak Family* both reflected and reinforced this view. And most of the newer family studies followed suit. They were generally preoccupied with the inheritance of mental defect. That was a major change from the time of Richard Dugdale, who identified only two cases of mental deficiency among the Jukes (attributing both to syphilis). In contrast, when Arthur Estabrook reanalyzed the Juke family in 1915, he judged half its members to be feebleminded.

Goddard Invents the Moron

Goddard himself was largely responsible for the new emphasis on mentality. On a trip to Europe in 1908, he had come across a test recently designed by the French psychologist Alfred Binet to detect mental deficiency in children. With Théodore Simon, Binet had also developed a system for classifying children according to their "mental age" (defined as the age of a corresponding group of children whose average test scores they matched). On his return to America, Goddard administered the Binet-Simon test to a wide range of children and found that the tests could be used to distinguish variations in the broad category of "feeblemindedness." He proposed a three-tiered system for grading mental deficiency. Those in the lowest grade, with a mental age of one or two, he termed "idiots," those in the middle grade, with a mental age of three to seven, "imbeciles," and those in the highest grade, with a mental age between eight and twelve, "morons". (The test was so hard at the upper end that more than half of those it rated morons would have been identified as normal by the Stanford-Binet test, which came into use in 1916 [Gelb 1987, 252].)

Goddard's system was actually grafted onto an earlier tradition of classification based on judgments of what constituted normal social behavior. The new language of mental defect was overlaid on an older concept of the moral imbecile, with deviance remaining a crucial mark of degeneracy. A feebleminded person, according to the standard definition, was "incapable of performing his duties as a member of society in the position of life to which he is born" (Popenoe 1915, 32). The psychologist Carl Brigham, who helped develop mental tests for the U.S. Army, explained that "the diagnosis of the border-line cases of feeblemindedness is, in the last analysis, a social diagnosis, and can not be based on intelligence tests alone" (1923, 145). That is why Elizabeth Kite could recognize feeblemindedness among the Kallikaks without benefit of mental tests.

In a series of lectures delivered at Princeton University, Goddard asked how the feebleminded could be distinguished from the "dull normal person." His answer: the former cannot tell right from wrong (1920, 87). This failure explains why the rate of feeblemindedness among criminals and other social misfits is so high. As Goddard had asserted in *Feeble-Mindedness*, "It is hereditary feeble-mindedness that is at the basis of all these [social] problems, and it is hereditary feeble-mindedness that we must attack and attack hard if we would solve them" (1914, 572). Thus behavior that in the nineteenth century would have been blamed on bad character was now taken to reflect a cognitive defect

(Gelb 1987, 249). Not everyone viewed feeblemindedness as the root cause of crime, alcoholism, prostitution, vagrancy, and pauperism. Charles Davenport, for instance, insisted that most antisocial behavior was the result of poor emotional control, a trait he did not think obviously correlated with intelligence (Davenport 1915a). Nonetheless, Goddard's influence was enormous. According to British sexologist Havelock Ellis, who praised Goddard's "careful" work, the feebleminded "form the reservoir from which the predatory classes are recruited" (1914, 34, 38). His view was echoed by Walter Fernald, a pioneer of American special education, who stressed that feeblemindedness was a cause of crime, prostitution, pauperism, and intemperance. "The feeble-minded are a parasitic, predatory class," he asserted. "They cause unutterable sorrow at home and are a menace and danger to the community. . . . Every feeble-minded person, especially the high-grade imbecile, is a potential criminal needing only the proper environment and opportunity for the development and expression of his criminal tendencies" (1912).

Goddard and Davenport did agree that misfits should not be blamed for their actions, since they could not control them. While most scientists believed that behavior was determined by heredity, the perceived cause of conduct is not what is crucial. Were actions determined by environment, the implications would be the same. Personal responsibility assumes choice; we do not punish behavior that is thought to be involuntary, as in cases of "temporary insanity." Thus, in the new view, the defectives' misdeeds were not their fault. Morons must certainly be segregated or (according to the older, more optimistic Goddard) appropriately trained. But to blame them for their behavior would be both futile and unjust, for they did not know that their actions were wrong (Davenport 1913; Goddard 1920, 81–87; Rogers and Merrill 1919, 347). (Of course, from the perspective of those designated morons, this might seem a distinction without a difference; in either view, the result was incarceration.)

Normal people, however, were expected to know right from wrong; for them, the story of the Kallikak family was meant to impart a moral lesson. Goddard preached: "Let the lesson be learned; let the sermons be preached; let it be impressed upon our young men of good family that they dare not step aside for even a moment" (1912, 103). Indeed, one historian has characterized the book as a "sermon of new science." She views the Kallikak story as a "parable" fusing the new principles of Mendelism to an older and still powerful Christian morality. Martin Kallikak, Sr. was the scion of a middle-class family of good English blood, which for four generations had maintained "a reputation for honor and respectability of which they are justly proud." But in an unguarded moment, he had sired an appalling line of mental defectives. Goddard's *The Kallikak Family* is a homily on the potentially disastrous genetic consequences of sowing wild oats. But the book's moralism helps account for the book's appeal to a deeply Christian audience (Zenderland 1992).

Galton had characterized eugenics as a "secular religion." Like many British intellectuals, he was in search of a satisfying substitute for Christianity. But even in Britain, Galton's open hostility to religion probably limited the popular appeal of his views. In the United States, associating eugenics with atheism would certainly have proved fatal. While Christianity was growing weaker in Europe, it was gaining strength in America; a larger proportion of Americans attended church on the eve of the Civil War than on the eve of the American Revolution (Butler 1990, 2). Indeed, that trend has continued to the present day. In the 1930s, 10 percent of Americans read the Bible daily; today it is 15 percent. More Americans believe in miracles, pray regularly, and attend worship services than do citizens of any other Western country (Gallup 1989, 45–48, 251). Given this religiosity, eugenicists have usually stressed the compatibility of Christianity with selective breeding. Thus Davenport wrote of the "grand work" done by religious teachers but argued that their ministrations are dependent on the nature of those who receive them. "Religion would be a more effective thing," he argued, "if everybody had a healthy emotional nature; and it can do nothing at all with natures that have not the elements of love, loyalty, and devotion" (1911, 255). Many others emphasized that eugenics and Christianity both abjured selfishness and asked individuals to sacrifice for a larger good. Goddard was especially adept at clothing eugenics in religious garb. He was able to impart scientific lessons in language that resonated with Americans.

The Mark of the Moron

One of those lessons was the hidden character of mental defect and thus the need for mechanisms of detection. Goddard explained that "a large proportion of those who are considered feeble-minded in this study are persons who would not be recognized as such by the untrained observer" (1912, 104). Indeed, the men were often physically robust, and the women, like Deborah Kallikak, were typically "good-looking, bright in appearance, with many attractive ways." For this reason teachers clung to the delusion "that such a girl will come out all right" (11–12). Goddard's frustration with parents and teachers was echoed by others who worked at state schools (Popenoe and Johnson 1918, 188).

So was his conviction that moronic women—mentally dull, oblivious to right and wrong, but often very pretty—were easy prey for immoral men, and, without supervision, would be lured into lives of waywardness.

Moreover, idiots and imbeciles, who "possess such inferior and ill-coordinated natural qualities that they require great bolstering up in order to reproduce at all," presented less threat than morons, who were ostensibly prolific. According to the 1914 report of the American Breeders' Association's Committee to Study and to Report on the Best Practical Means of Cutting Off the Defective Germ-Plasm in the American Population, "It is the moron or high-grade feeble-minded class of individuals that constitute the greatest cacogenic menace, for these individuals . . . are able to, and do reproduce their unworthy kind (Laughlin 1914, 19). That view was echoed by the authors of "Dwellers in the Vale of Siddem," who explained:

> It is not the idiot or, to any great extent, the low grade imbecile, who is dangerous to society. In his own deplorable condition and its customarily accompanying stigmata, he is sufficiently anti-social to protect both himself and society from the results of that condition. But from the high grade feeble-minded, the morons, are recruited the ne'er-do-wells, who . . . drift from failure to failure, spending a winter in the poor house, moving from shack to hovel and succeeding only in the reproduction of ill-nurtured, ill-kempt gutter brats to carry on the family traditions of dirt, disease, and degeneracy. (Rogers and Merrill 1919, 346–47).

Numerous statistical studies, in both Britain and America, seemed to confirm these alarms. Thus the British Royal Commission on the Care and Control of the Feeble-Minded reported that defectives averaged seven children, normal couples only four (1908, 29). William Hutton, director of the Eugenics Society of Canada, reported that Canadian families listed in *Who's Who* produced an average of 2.42 children, whereas the parents of inmates at the Orilla Asylum for the Feeble-Minded had 8.7 (McLaren 1990, 115). Virtually every commentator explained the imbalance as a consequence of feebleminded persons' lack of self-restraint; in the words of Leonard Darwin, secretary of the British Eugenics Education Society, defectives "have large families and many descendants because they have little power of looking into the future, or of foreseeing the consequence of their own acts" (1928, 46–47). That the feebleminded bred with no regard for consequences was a central theme of the later family studies. Richard Dugdale had assumed that degenerate lines would naturally become extinct over time. His successors assumed the opposite: that they would swamp normal families.

A decline in the birthrate of college graduates prompted special concern. After reviewing many studies, the distinguished American psychologist G. Stanley Hall concluded, "If higher education became universal, posterity would gradually be eliminated and the race progressively exterminated by schools and teachers" (1904, 608). It appeared that nearly a quarter of Harvard and Yale men never married, while the remainder produced a mere 1.9 surviving children each. At that rate, they would not even maintain the stock. Statistics from the women's colleges were even worse. Only half the graduates of Wellesley married, while the mothers averaged 1.56 children each (Phillips 1916).

Although the low reproductive rates of college graduates probably was of great concern only to professionals, the fecundity of the feebleminded was a different matter. Graduates of elite colleges numbered in the hundreds, mental defectives in the hundreds of thousands. Exactly how many, no one knew, but it was commonly estimated that from 300,000 to 1,000,000 persons were feebleminded as a result of genetic defect. Some commentators believed the figure much higher. A few thought that society might benefit from the presence of more high-grade feebleminded. They reasoned that a growing country has need for labor that does not appeal to the intelligent, that it takes all kinds to do the world's work. According to distinguished sociologist Ferdinand Tönnies, since most people necessarily did work of "hopeless monotony," their mental dullness was an asset (1906, 40–41). But many more feared being swamped by a rising tide of mental defectives. Davenport attributed the rise of the eugenics movement to belief in the "great proportional increase in feeble-mindedness in its protean forms" (1912, 308), while Justice Oliver Wendell Holmes (in the 1927 case *Buck v. Bell*) asserted a need "to prevent our being swamped with incompetence." A raft of books, articles, and reports agonized over the problem of feeblemindedness, as did myriad government agencies, public investigating commissions, and private committees. In 1900, mental deficiency was considered an insignificant problem. By 1915, in contrast, it was seen "as perhaps the largest and most serious social problem of the time" (Davies 1930, 94).

Estimates of the number of mental defectives tended to increase as intelligence tests came into ever wider use—to evaluate students, prisoners, inmates of poorhouses and schools for the feebleminded, immigrants at Ellis Island, and army draftees. Prior to the introduction of Binet testing, a common estimate was that about five individuals in a thousand were feebleminded. In 1912, Goddard tested New York City schoolchildren and estimated that 2 percent were probably feebleminded (Goddard 1911–12). The sense of alarm engendered by

FIGURE 4.3   "No Race Suicide in This Allegheny College Family." Frontispiece from H. R. Hunt, "The Allegheny College Birthrate," *Journal of Heredity* 14 (May 1923): 51–60. Photo courtesy of Marine Biological Laboratory, Woods Hole, Massachusetts.

such statistics is reflected in a passage from a textbook authored by five leading geneticists:

> The proportion of those who are feeble-minded in such various directions as to constitute the feeble-minded class is estimated at 3 per cent of our population, and were we to include drunkards, paupers, grave sex-offenders, the criminalistic, the insane, and those with innate physical weaknesses that render them for the most part incompetent, it seems a safe estimate that 8 per cent of our population are far from having the capacities of effective men and women, able, not merely to support themselves, but really to push forward the world's work. (Castle et al. 1912, 308–9)

The Army Tests Its Recruits

Alarm turned to panic after the First World War, as the results of the U.S. Army testing program were widely disseminated. Robert Yerkes, then president of the American Psychological Association, developed the tests in collaboration with Goddard and other psychologists, initially as a means for rejecting the mentally unfit. But the psychologists, seizing an opportunity to advance their profession, turned it into a much more ambitious program for determining service assignments and selection for officer training. (Charles Davenport also proposed a scheme for choosing naval officers based on their pedigrees [Davenport 1917b].) Many regular army officers were resentful, believing that they were perfectly competent to judge recruits' capabilities, which they saw as much more closely linked to character than to intelligence (Carson 1993; Kevles 1968–69). The army had valued cheerfulness, loyalty, and reliability in its recruits, resourcefulness, initiative, tact, and leadership ability in its officers. Prior to the war, explicit assessments of officers' intelligence were extremely rare. By the war's end, such assessments had become a major criterion in the reports that officers prepared on each other and their subordinates (Carson 1993, 281).

Ultimately, more than 1.75 million soldiers took the army Alpha test, if they were literate in English, or the army Beta test (which used pictures), if they were not. Yerkes claimed that the tests measured native intelligence, not education or training. Carl C. Brigham, who helped prepare the exam, noted in his popular book *A Study of American Intelligence* that the information test had been heavily criticized but countered that "the assumption underlying the use of a test of this type is that the more intelligent person has a broader range of general information than an unintelligent person. Our evidence shows that this assumption is, in the main, correct" (1923, 31).

PSYCHOLOGICAL EXAMINING IN THE UNITED STATES ARMY

TEST 8

Notice the sample sentence:

People hear with the   eyes   <u>ears</u>   nose   mouth

The correct word is ears, because it makes the truest sentence.

In each of the sentences below you have four choices for the last word. Only one of them is correct. In each sentence draw a line under the one of these four words which makes the truest sentence. If you can not be sure, guess. The two samples are already marked as they should be.

SAMPLES { People hear with the   eyes   <u>ears</u>   nose   mouth
{ France is in   <u>Europe</u>   Asia   Africa   Australia

1   America was discovered by   Drake   Hudson   <u>Columbus</u>   Cabot...................   1
2   Pinochle is played with   rackets   <u>cards</u>   pins   dice.........................   2
3   The most prominent industry of Detroit is   <u>automobiles</u>   brewing   flour   packing......   3
4   The Wyandotte is a kind of   horse   <u>fowl</u>   cattle   granite...................   4
5   The U. S. School for Army Officers is at   Annapolis   <u>West Point</u>   New Haven   Ithaca..   5

6   Food products are made by   Smith & Wesson   <u>Swift & Co.</u>   W. L. Douglas   B. T. Babbitt   6
7   Bud Fisher is famous as an   actor   author   baseball player   <u>comic artist</u>...........   7
8   The Guernsey is a kind of   horse   goat   sheep   <u>cow</u>...........................   8
9   Marguerite Clark is known as a   suffragist   singer   <u>movie actress</u>   writer.........   9
10   "Hasn't scratched yet" is used in advertising a   duster   flour   brush   <u>cleanser</u>........   10

11   Salsify is a kind of   snake   fish   lizard   <u>vegetable</u>.........................   11
12   Coral is obtained from   mines   elephants   oysters   <u>reefs</u>.......................   12
13   Rosa Bonheur is famous as a   poet   <u>painter</u>   composer   sculptor...................   13
14   The tuna is a kind of   <u>fish</u>   bird   reptile   insect..........................   14
15   Emeralds are usually   red   blue   <u>green</u>   yellow...........................   15

16   Maize is a kind of   <u>corn</u>   hay   oats   rice.............................   16
17   Nabisco is a   patent medicine   disinfectant   <u>food product</u>   tooth paste..........   17
18   Velvet Joe appears in advertisements of   tooth powder   dry goods   <u>tobacco</u>   soap.....   18
19   Cypress is a kind of   machine   food   <u>tree</u>   fabric.........................   19
20   Bombay is a city in   China   Egypt   <u>India</u>   Japan.........................   20

21   The dictaphone is a kind of   typewriter   multigraph   <u>phonograph</u>   adding machine......   21
22   The pancreas is in the   <u>abdomen</u>   head   shoulder   neck.....................   22
23   Cheviot is the name of a   <u>fabric</u>   drink   dance   food.....................   23
24   Larceny is a term used in   medicine   theology   <u>law</u>   pedagogy..................   24
25   The Battle of Gettysburg was fought in   <u>1863</u>   1813   1778   1812.................   25

26   The bassoon is used in   <u>music</u>   stenography   book-binding   lithography..........   26
27   Turpentine comes from   <u>petroleum</u>   ore   hides   trees.....................   27
28   The number of a Zulu's legs is   <u>two</u>   four   six   eight....................   28
29   The scimitar is a kind of   musket   cannon   pistol   <u>sword</u>...................   29
30   The Knight engine is used in the   Packard   Lozier   <u>Stearns</u>   Pierce Arrow...........   30

31   The author of "The Raven" is   Stevenson   Kipling   Hawthorne   <u>Poe</u>.................   31
32   Spare is a term used in   <u>bowling</u>   football   tennis   hockey..................   32
33   A six-sided figure is called a   scholium   parallelogram   <u>hexagon</u>   trapezium.........   33
34   Isaac Pitman was most famous in   physics   <u>shorthand</u>   railroading   electricity........   34
35   The ampere is used in measuring   wind power   <u>electricity</u>   water power   rainfall......   35

36   The Overland car is made in   Buffalo   Detroit   Flint   <u>Toledo</u>.................   36
37   Mauve is the name of a   drink   <u>color</u>   fabric   food......................   37
38   The stanchion is used in   fishing   hunting   <u>farming</u>   motoring................   38
39   Mica is a   vegetable   <u>mineral</u>   gas   liquid..........................   39
40   Scrooge appears in   Vanity Fair   <u>The Christmas Carol</u>   Romola   Henry IV...........   40

FIGURE 4.4   Sample page from U.S. Army mental tests. Examination Alpha, Test 8: Information. Forms 8 and 9. From Robert M. Yerkes, ed. *Psychological Examining in the United States Army*, vol. 15 of *Memoirs of the National Academy of Sciences* (Washington, D.C., 1921).

Although the army program ended with the war, its influence was vast. Previously, tests had been directed at schoolchildren and inmates of custodial institutions. Moreover, the results had to be interpreted by trained professionals. The army tests could be rapidly administered to the general population and produced a quantified, apparently objective, result. They demonstrated that mass testing was practical. As a direct consequence, the use of tests in education and business vastly expanded. At the same time, they greatly increased anxiety about the level of national intelligence, for the army tests ostensibly demonstrated that nearly half of the white draft (47.3 percent) was feebleminded (Brigham 1923, 80–86; Yerkes 1921, 785). The lowest scores were registered by blacks and members of the newer immigrant groups (a subject discussed in more detail in chapter 6). These results were publicized in scores of books and articles lamenting the sorry state of Americans' mentality and warning of its further decline. But in the perspective of many geneticists, it would come to seem that both the extent of the problem and the difficulty of finding a solution were actually greater than they had appreciated. The reason lies in an increasing awareness of the implications of Mendelism.

The "Real Menace" of the Feebleminded

In *The Kallikak Family*, Goddard provided only a cursory discussion of Mendelism, asking readers to wait for a larger book. *Feeble-Mindedness: Its Causes and Consequences* appeared two years later in 1914 and explicitly discussed the meaning of the Kallikak data for theories of inheritance. According to Goddard, normal-mindedness was a dominant trait, transmitted in accordance with Mendel's laws. Thus individuals possessing at least one copy of the normal factor for intelligence would be normal, while those lacking both copies would be mentally defective. Not everyone thought that the genetics of mentality was so straightforward. Thus Davenport recognized that "feeble-mindedness is not a biological, but a social term" (Davenport 1915c, 837). (If feeblemindedness were socially defined, one would not expect it to be transmitted as a simple Mendelian recessive.) Nonetheless, Goddard was in general successful in convincing geneticists that much antisocial behavior was attributable to recessive genes (Barker 1989). That principle had implications that Goddard did not foresee and could hardly applaud. For if these bad genes were recessive, programs of eugenic segregation and sterilization would not be nearly as efficient as their proponents had claimed.

In *The Kallikak Family*, Goddard had asserted that segregation would reduce the number of feebleminded from 300,000 to 100,000 "in a single generation" (1912, 106). The view that mental defect could be rapidly reduced was very widely shared. Thus Harry Laughlin claimed that "it would be possible at one fell stroke [to] cut off practically all of the cacogenic varieties of the race" (1914, 47; see also 60). Herbert Eugene Walter argued that "could such a [sterilization] law be enforced in the whole United States, less than four generations would eliminate 9/10 of the crime, insanity, and sickness of the present generation in our land. Asylums, prisons, and hospitals would decrease, and the problem of the unemployed, the indigent old, and the hopelessly degenerate would cease to trouble civilization" (1913, 255). Florence Mateer (of the Vineland Training School) asserted that if procreation were stopped, the burden of supporting defectives "would be practically under control in a generation" (1913, 227). Charles Davenport likewise argued that a policy of segregating defectives during their reproductive years would—after one generation—reduce their numbers "to practically nothing" (1912, 286).

But if mental defects were attributable to *recessive* genes, all these predictions were wrong. The Harvard geneticist Edward East was the first to see why. In 1917, East showed that if feeblemindedness were transmitted as a Mendelian recessive, the number of apparently normal carriers must be vastly larger than those affected. These heterozygotes would not be touched by programs of eugenical segregation and sterilization. The "real menace" of the feebleminded, he argued, lay in the mass of invisible carriers, which constituted about 7 percent of the American population, or one in every 14 individuals. "Our modern Red Cross Knights," East exclaimed, "have glimpsed but the face of the dragon" (1917, 215).

The problem of the high-grade feebleminded paled in comparison. At least they could be detected by mental tests or well-trained observers. The carriers would not even themselves know that they were "tainted." As a result, they could unwittingly contaminate others. Thus feeblemindedness came to be viewed as a kind of infectious disease, which could be unknowingly spread. H. S. Jennings compared the heterozygotes to "the carriers of the typhoid bacillus that are themselves immune to the disease. Though themselves unaffected, they hand on the source of ill to others" (1930, 234).

If most defective genes were hidden in apparently normal carriers, eugenic measures would work much more slowly than anyone had thought. Applying the Hardy-Weinberg formula (which allows one to calculate the frequency of heterozygote carriers when the frequency of the gene is known), the British geneticist R. C. Punnett derived figures even more alarming than East's: over 10 percent of the population apparently carried the defective gene (Punnett 1917). On the unrealistic assumption that all the affected could be prevented from breeding, it would take over 8,000 years before their numbers were reduced to one in 100,000. Punnett had once thought that a policy of strict segregation would rapidly eliminate feeblemindedness. He now concluded that eugenic segregation did not, after all, offer a hopeful prospect. To achieve sufficient progress, he argued, it would be necessary to identify carriers. That was where he thought research should be directed.

Punnett's figures were seized on by opponents of eugenics, who argued that segregation and sterilization of the affected worked too slowly to justify the effort. Neither Punnett nor his geneticist colleagues drew that conclusion. Indeed, Edwin G. Conklin asserted that "all modern geneticists approve the segregation or sterilization of those who are known to have serious hereditary defects, such as hereditary feeblemindedness, insanity, etc." (1930, 577). That was only a slight exaggeration. These geneticists certainly understood that policies of eugenical segregation and sterilization would work very slowly. According to J. H. Kempton, even the most ardent advocates of sterilization and segregation "admit that its beneficial effects are slow and slight" (1933, 465). That was certainly true of H. S. Jennings, who conceded that "to merely cancel the deficient individuals themselves— those actually feebleminded—makes almost no progress toward getting rid of feeblemindedness for later generations" (1927, 273). But Jennings did not therefore conclude that the policy was wrong. On the contrary, he claimed that allowing the feebleminded to propagate was a "crime." Today it is often said that to grasp the implications of the Hardy-Weinberg formula is to recognize the futility of eugenics. But the geneticists who first articulated these implications did not see things that way. Why not?

First, they believed that every case saved was a gain and would therefore approve eugenic policies whatever their exact effect (Jennings 1930, 238). Jennings wrote: "A defective gene—such a thing as produces diabetes, cretinism, feeblemindedness — is a frightful thing; it is the embodiment, the material realization of a demon of evil; a living self-perpetuating creature, invisible, impalpable, that blasts a human being in bud or leaf. Such a thing must be stopped wherever it is recognized" (1927, 274). That view was echoed by many others, including Lancelot Hogben, who argued that the fact that we cannot do everything "is not a valid reason for neglecting to do what little can be done" (1931, 207), and Curt Stern, who asserted that the fact that selection against "severe physical and mental abnormalities will reduce the number of affected from one generation to the next by only a few per cent does not alter the fact that these few per cent may mean tens of thousands of unfortunate individuals who, if never born, will be saved untold sorrow" (1949, 538).

In fact, nearly all geneticists of the 1920s and 1930s—including those traditionally characterized as opponents of eugenics (which includes Hogben, Jennings, and Conklin)—took for granted that the feebleminded should be prevented from breeding. Many would have approved segregation or sterilization even if they were convinced that mental defect was not heritable. They thought such persons would make incompetent parents; a view that the public generally shared (as we will see in the next chapter). Indeed, as the genetic case for controls grew increasingly weak, it was largely superseded by a social one.

Modern Views

In 1930, Conklin asked rhetorically how any serious objection could be raised to the American Eugenics Society's proposed sterilization policy (577–78). Today, references to that same policy are invariably derisive. Indeed, between 1930 and the present, a dramatic change occurred in attitudes toward controlled reproduction. This shift in opinion cannot be explained by the progress of genetics. It is sometimes said that eugenics was shown to be futile when it was finally understood that most of the offending genes were carried in apparently normal individuals in whom they are not expressed. But we have seen that this point was not only appreciated after 1917 by many eugenicists but employed by them to argue for the necessity of methods to detect carriers. For most geneticists, its implication was that eugenic programs should be widened, not abandoned.

It is true that much of Goddard's work has been discredited. His data on the incidence of mental defect are now considered worthless, as is the rubric of feeblemindedness. But these developments do not explain the much later reversal in attitudes. Already in the 1910s, Davenport had characterized feeblemindedness as a lumber room in which disparate traits were thrown together and noted that it was a social rather than biological category. Objections to methods of collecting family-study data and to the view that even true mental defect could be a simple recessive trait were frequently raised in the 1920s. Many of these critics were other eugenicists, who strongly supported policies to prevent those they considered genuine mental defectives from breeding. Moreover, in the United States, Canada, and Scandinavian countries, as well as in Germany, eugenics found its widest application in the 1930s—long after the scientific events that are often cited to explain its decline.

In 1918, Popenoe and Johnson wrote that "so few people would now contend that two feeble-minded or epileptic persons have any 'right' to marry and perpetuate their kind, that it is hardly worth while to argue the point" (170). Today, in contrast, few would dare defend it. Assumptions of reproductive autonomy we now take for granted they thought too absurd even to require challenging. The inversion of these assumptions in recent decades is primarily explained by political developments. Revelations of Nazi atrocities, the trend toward respect for patients' rights in medicine, and the rise of feminism have converged to make reproductive autonomy a dominant value in our culture. In 1914, a committee of the American Breeders' Association asserted that "society must look upon germ-plasm as belonging to society and not solely to the individual who carries it" (Laughlin 1914, 16). Almost no one today would profess such a belief. Indeed, the dominant view is now its opposite: that the nature of reproductive decisions should be no concern of the state. In the next three chapters we will explore the events that led to that sea-change in attitudes.

Eugenic Solutions

A LL EUGENICISTS SHARED at least one conviction: that reproductive decisions should be guided by social concerns. On virtually every other question, they divided. Who should be urged to breed: college graduates, the middle class, "Nordics"? Who should be discouraged from breeding: workers, the chronically poor, mental defectives, immigrants? Who should make these decisions: individual families, social organizations, the state? What forms should eugenics policy take: education and moral suasion, tax rebates and family allowances, the wider availability of birth control, segregation in custodial institutions, sterilization?

In no country was there consensus on any of these questions. But support for more stringent measures increased everywhere as economic conditions deteriorated. The effect of the world economic crisis of the 1930s is particularly evident with respect to sterilization: a host of new enabling laws were passed, and old ones were strengthened and applied more vigorously. In the United States, the number of sterilizations climbed (Reilly 1991, 93–102). Sterilization was legalized in Germany (1933), the Canadian province of British Columbia (1933), Norway (1934), Sweden (1934), Finland (1935), Estonia (1936), and Iceland (1938). Denmark, which in 1929 had legalized "voluntary" sterilization, passed a new statute permitting its coercive use on mental defectives in 1934. The Canadian province of Alberta, which passed a sterilization law in 1928, removed the requirement for voluntary consent in 1937.

However, there were important national differences in attitudes toward eugenics. In Britain, coercive sterilization was never seriously considered. Most British eugenicists hoped their goals would be met by persuading people that they had a responsibility to the "race" and providing them with wider access to birth control. They thought that people would be convinced through education to exercise reproductive restraint voluntarily. (Some eugenicists were less sanguine, but they were resigned to the view that compulsory measures were politically impractical.)

Hereditarian beliefs were certainly strong in Britain, and much of the public apparently favored sterilization of the mentally handicapped (Radford 1991, 456). But these beliefs did not translate into a legislative program. Effective opposition from organized labor, as well as from Catholics, was a principal reason. The British working class was relatively unified, whereas in America it was fractured not only by ethnicity and religion—which in the United States were often the most important determinants of party choice—but by workers' geographical mobility (McCormick 1990, 96, 100). Eugenicists in Germany and America were able to play on racialist fears that cut across class lines; eugenics was often aimed at "outsiders." In Germany, Jews were damned; in the United States and Canada, the foreign-born. But reflecting Britain's sharp class divisions, the target of eugenics in that country was almost exclusively the urban poor (Mazumdar 1992). Not surprisingly, they resented it. And unlike Jews and recent immigrants, they were politically powerful.

British eugenicists tried to distinguish the respectable working class from the chronically poor. The latter group, variously known as the "submerged tenth," the "social problem group," or the "social residuum" (reformulating an old distinction between the worthy and the unworthy poor), was held responsible for every social ill that beset the city slums. But in practice, the distinction between workers and their residuum was often blurred, as in the many reports lamenting the high birthrate of the lower classes. The historian G. R. Searle has remarked that one of the most striking features of the eugenics literature is the way in which working-class people are often "discussed as though they were denizens of some other planet" (1976, 60).

While a few members of the Labour Party in Parliament promoted eugenics, on the whole labor representatives saw it as a diversion from more important programs of social reform (Jones 1986, 62). They also suspected that the eugenicists' program was directed at workers. That they were right became increasingly clear as the economic crisis deepened. For example, the biologist Julian Huxley urged in 1931 that individuals receiving unemployment benefits be barred from reproducing. "Infringement of this order could probably be met by a short period of segregation in a labour camp," he wrote. "After three or six months' separation from his wife he [an unemployed man] would be likely to be more careful the next time" (quoted in Werskey 1978, 42). And Huxley was considered a very moderate eugenicist.

In Britain as in other countries, eugenics was opposed by Catholics of all political persuasions and by those "classical" liberals who believed that the functions of the state should be kept to an absolute minimum. In Britain, eugenics was effectively opposed by the organized labor movement as well. Given the political strength of the British working class, the various eugenics organizations had to tread very carefully. In appealing to a wider public, the Eugenics Education Society had constantly to deny that its policies were informed by class prejudice or would threaten personal liberties. It made no effort to pass a compulsory sterilization law. Indeed, it failed even in efforts to legalize voluntary sterilization (which Catholics opposed on principle and labor suspected would be coercive in practice). The Mental Deficiency Act of 1913 allowed for eugenically motivated segregation. But despite intense efforts by the eugenics society, reinforced by the 1934 recommendation of the Brock Committee that voluntary sterilization be offiicially sanctioned, the 1913 act with its emphasis on segregation remained in effect until 1959 (Radford 1991, 452). Even efforts to segregate were relatively modest. American institutions housed many persons who would not have been certifiable as defective in Britain.

Relative to the situation in the United States, eugenics in Britain was constrained in financial, structural, and theoretical respects as well. The movement enjoyed some philanthropic support, but on a much smaller scale than in the United States. In Britain, all laws required parliamentary approval, whereas Americans could target the most sympathetic state legislatures. Moreover, at the state level, small well-organized groups or even a few activists could be politically effective. In a number of American states, a handful of people lobbying key legislators was responsible for the passage of sterilization laws. Thus the Chairman of the the American Breeders' Association Committee on Defective Germ-Plasm reported that "the laws already enacted have usually been put through by some very small energetic group of enthusiasts, who have had influence in the legislatures. In at least two of the states it was chiefly the work of a physician. In one, of a woman" (van Wagenen 1912, 477). That pattern continued into the 1920s and 1930s. For example, in Georgia, the Augusta Junior League in conjunction with a handful of mental health experts and physicians was responsible for passage of the last American law in 1937 (Larson 1991).

British geneticists were also a very fractious group. Thus Karl Pearson refused to have anything to do with the Eugenics Education Society, whose work he considered sloppy. That geneticists were divided into warring camps of Mendelians and biometricians (with the society dominated by the former) undermined their authority, notwithstanding agreement on the value of eugenics.

Eugenics did enjoy support from some socialists, most notably members of the Fabian Society, who rejected laissez-faire in favor of a planned economy and establishment of a "National Minimum"—a guaranteed level of health, education, wages, and employment. Founded in 1900, the society was committed to reform from above rather than revolution from below. While the Fabians condemned capitalism, their ideal was a scientifically planned society that would empower experts rather than workers. The Fabians envisioned a nation managed by people much like themselves: middle-class professionals, such as doctors, scientists, teachers, and social workers. The society also attracted a number of important literary figures, including George Bernard Shaw, who believed that "there is now no reasonable excuse for refusing to face the fact that nothing but a eugenic religion can save our civilisation," and H. G. Wells, who argued for "the sterilization of failures" on the grounds that "the way of Nature has always been to slay the hindmost, and there is still no other way, unless we can prevent those who would become the hindmost being born" (Wells 1905, 60; Shaw 1905, 74). Plays like Shaw's *Man and Superman* (1903) and novels like Wells's *A Modern Utopia* (1905) probably did more than any academic studies to popularize the concept of selective breeding.

Fabian socialism and eugenics shared the conviction that laissez-faire was a bankrupt philosophy that should be replaced by planning based on social needs. The geneticist Lancelot Hogben noted: "Negative eugenics is simply the adoption of a national minimum of parenthood, and extension of the principle of national minima familiarized in the writings of Sidney and Beatrice Webb. It is thus essentially *en rapport* with the social theory of the collectivist movement" (1931, 210). Sidney Webb himself enthusiastically endorsed the claim: "No consistent eugenist can be a 'Laisser Faire' individualist unless he throws up the game in despair. He must interfere, interfere, interfere!" (1910–11, 237).

Fabians were often nationalist and imperialist, though few went as far as Wells, who bluntly asserted that "there is only one sane and logical thing to be done with a really inferior race, and that is to exterminate it" (quoted in Trombley 1988, 32; see also Coren 1993, 65–67). But they vacillated in their attitudes toward the poor, viewing them sometimes with sympathy, sometimes with contempt. The tension between two images of the poor—as exploited and as unfit—was reflected in the Fabian political program, which promoted eugenics simultaneously with measures for improved health, education, and welfare (Kramnick and Sheerman 1993, 37).

Fabians tried to resolve this tension by strongly distinguishing the prudent working class from its residuum. They were as alarmed as conservatives at the purported fecundity of the lower classes. Thus the Fabian theorist Harold Laski (who worked briefly in Karl Pearson's laboratory) warned that the unfit were outbreeding the fit and that society must learn to regard "the production of a weakling as a crime against itself" if it were not to commit race suicide (1910, 25–34; see also Kramnick and Sheerman 1993, 30–48). Many others agreed. Laski's alarmist view was echoed by Eden Paul: "Unless the socialist is a eugenicist as well, the socialist state will speedily perish from racial degradation" (1917, 139), while Wells asserted that "we cannot go on giving you health, freedom, enlargement, limitless wealth, if all our gifts to you are to be swamped by an indiscriminate torrent of progeny" (1922, xvi).

However, there was often a large disparity between the Fabians' extreme rhetoric and their milder policy proposals, which rarely extended beyond segregation. Very few (other than Wells) supported coercive sterilization. Thus Havelock Ellis invoked terrifying images: "When we are able to control the stream at its source we are able to some extent to prevent the contamination of that stream by filth, and ensure that its muddy floods shall not sweep away the results of our laborious work on the banks." But he repeatedly rejected coercive sterilization (Ellis 1914, 15–16, 30). In Ellis's view, eugenics would be effective only if developed from a broader sense of social responsibility. Harold Laski had argued that any action with national consequences may be regulated by the state and urged that the unfit be prevented from breeding. But when it came to public policy, he asked only that the state influence the climate of public opinion. In the end, Laski would counter the threat of race suicide with education. In Britain, that was the best eugenicists could do.

Their inability to pass laws may lead us to dismiss the British eugenicists' importance. But legislation is not the only —or perhaps even the best—measure of success. The eugenicists were extremely effective in popularizing a new Galtonian vocabulary. Whole sections of British society now took for granted that talent and character were inborn and fixed. Edgar Schuster and Ethel Elderton of the Galton Laboratory remarked, "At the time of the first publication of Mr Galton's *Hereditary Genius*, in 1869, the belief in the hereditary nature of inborn natural ability was held by very few; but so great has been the influence of that and other works that at the present time it would be almost impossible to find an educated person to dispute it" (1907, 1). This assumption had consequences far beyond programs of eugenical selection. It shaped policy in respect to medicine, law, and education.

In the United States, eugenicists did enjoy some legislative triumphs, although even here the greatest impact was probably ideological. As many scholars have noted, eugenics was congruent with the scientific and reformist spirit of the Progressive Era, a period of vast economic and social change between the collapse of Reconstruction and the start of the First World War (Allen 1989; Freeden 1979; Pickens 1968). When the Civil War ended in 1865, the United States was an agricultural country, which imported most of its technology. By the end of the century, its industrial output had tripled, with steel production exceeding the combined output of Britain and Germany (Painter 1987, xvii). The United States was now a major exporter of industrial equipment and consumer goods. Business and industry became highly consolidated, as small-scale competition gave way to a new system of corporate capitalism in which most sectors of the economy were dominated by a few giant firms. At the same time, the population became increasingly urban. In 1880, about a quarter of Americans lived in cities; by 1900, the figure was 40 percent. These cities now filled with immigrants from Europe and poor migrants from the rural South while middle-class residents moved to new "streetcar suburbs."

The wealth so visibly created was very unequally distributed. At the turn of the century, the average workweek was about 60 hours, and the conditions in mines, mills, and factories were wretched; industrial discipline was harsh, and the work was exhausting and often dangerous. Widespread unemployment accompanied frequent and sometimes prolonged depressions. Signs of social disorder were everywhere: in strikes and walkouts that often ended in violence, in highly visible urban slums, in municipal corruption, in rising rates of crime, prostitution, alcoholism, and infectious diseases, in overcrowded prisons, and in asylums for the insane and feebleminded. The middle class demanded reforms that would both relieve distress and restore a stable social order. They called for factory inspection, child labor laws, a shortened workday, community clinics, probation and parole, a federal income tax, workmen's compensation, the direct election of U.S. senators, prohibition. And eugenics.

The Progressive reforms involved a vast expansion in governmental authority. Whether Democrat or Republican, the Progressives shared a faith in the virtues of planning and the benevolence of the state. Their bywords were "organization," "cooperation," "systematic planning," "efficiency," and "social control." Julia Lathrop, Hull House resident and later chief of the Children's Bureau, summarized their credo: "The success of our future civilization lies in government adding to their responsibility and taking on work which people have not hitherto been willing to entrust to them" (quoted in Rothman 1980, 6). That work would be the province of disinterested experts.

To the Progressives, science provided a model of impartial expertise. Moreover, science, being disinterested, could provide unity to a society that seemed to be culturally disintegrating. Above all, science could supply the tools to manage humans and their environment efficiently (Tobey 1971, 12–19). Science could also address the root causes of social problems and not just their symptoms. Like the state, it was assumed to be wholly benevolent. The Progressive attitude was expressed by Charles R. Van Hise, president of the University of Wisconsin:

> We know enough about agriculture so that the agricultural production of the country could be doubled if the knowledge were applied. We know enough about disease so that if the knowledge were utilized, infectious and contagious diseases would be substantially destroyed in the United States within a score of years; we know enough about eugenics so that if the knowledge were applied, the defective classes would disappear within a generation. (quoted in Haller 1985, 76)

Van Hise was also active in the conservation movement, as were a number of prominent eugenicists such as Theodore Roosevelt, Gifford Pinchot, Madison Grant, and Charles Goethe. Both movements emphasized the need for planning and the welfare of future generations. Progressives, like Fabians (whom they resembled in many respects), vacillated between sympathy and contempt for the poor, supporting measures both to ameliorate their plight and to prevent them from breeding. Like the Fabians also, they tried to resolve the tension by distinguishing workers from what in America were called the "defective" or "dangerous classes" (those in Britain called "the social residuum").

The financial burden of caring for the defectives appeared to be mounting. Large-scale custodial institutions—prisons, reformatories, training schools, insane asylums—first came into being in the 1820s, during the Jacksonian era. In the latter half of the nineteenth century, their number and scale rapidly expanded. Proportionally, the increase in institutional places far exceeded the growth of population (Wolfensberger 1975, 88). Charles Davenport spoke for many when he wrote: "It is a reproach to our intelligence that we as a people . . . should have to support about half a million insane, feeble-minded, epileptic, blind and deaf, 80,000 prisoners and 100,000 paupers at a cost of over 100 million dollars per year" (1911, 4).

As the asylums expanded, their conditions deteriorated. Although originally created as a substitute for brutal methods of dealing with criminals, the delinquent, and the insane, custodial care in practice was rarely humane. Cruelty and corruption came to seem more acceptable as the asylums filled with the poor and foreign-born. In 1890, 40 percent of inmates in state mental institutions were immigrants or the children of immigrants; in industrial regions, the percentage was even higher. Whether they were of old American or recent foreign stock, the inmates were mostly indigents and paupers. Under these circumstances, the sordid conditions did not seem so shocking (Rothman 1980, 17–24; see also Trent 1994, 167).

As the number of inmates rose, cost became an increasingly serious concern. The custodial institutions were expensive to run and came to account for a substantial proportion of state budgets. In Pennsylvania, the cost of operating asylums increased by 35 percent in the single decade of the 1880s (Reilly 1991, 17). Legislators were moved by fears of social disorder to appropriate funds for construction, but they had little incentive to pay for maintenance once inmates were warehoused. The result was constant pressure on operating costs, which in turn led to reduced training programs and physical comforts (Radford 1991). Legislators, like superintendents of state institutions, reasoned that inmates could work very hard (sometimes justifying their labor as training) and that most came from poor families and were not used to living in much comfort (Wolfensberger 1975, 49).

The "colony system" of organizing such institutions extended the principle of self-sufficiency. Large-scale farms, designed to yield produce for both the institution and market, were established on sites some distance from their parent institution. In his article "Waste Land plus Waste Humanity," Superintendent E. R. Johnstone of the Vineland Training School, a leading figure in the movement, explained its theory: "The colony should be located on rough uncleared land—preferably forestry land. Here these unskilled fellows find happy and useful occupation, waste humanity taking waste land and thus not only contributing toward their own support, but also making over land that would otherwise be useless." The principle could also be extended to females. According to Johnstone, the "girl-women" could successfully grow fruits and vegetables, raise poultry, pickle and can, sew and knit. "No manufacturer of to-day has let the product of his plant go to waste as society has wasted the energies of this by-product of humanity," he wrote (quoted in Popenoe and Johnson 1918, 189).

However, the colonies ultimately failed to achieve self-sufficiency. The founders of the first colonies had chosen especially good farms and the most capable residents to operate them; as the system expanded, its efficiency declined (Davies 1930, 226). Paul Popenoe and Roswell Johnson, authors of the popular textbook *Applied Eugenics* (1918), defended the expansion of the segregation system along the Vineland model. Popenoe and Johnson claimed that costs would actually decline,

FIGURE 5.1 "Feeble-Minded Men Are Capable of Much Rough Labor." From Paul Popenoe and Roswell H. Johnson, *Applied Eugenics* (New York: Macmillan, 1918), p. 192. Courtesy of Francis Countway Medical Library, Harvard University.

since a program of sexual quarantine would continually decrease the number of wards of the state. They also noted that institutions for the feebleminded were run at a much lower per capita cost than were prisons or insane asylums. On the assumption that many prison and mental asylum inmates were also feebleminded, Popenoe and Johnson argued that the state could save money by transferring inmates to the less expensive institutions. They noted as well that a large part of the expense could be met by better organization of inmate labor. But costs continued to mount, and so did doubts about the prospects for cure. That combination prompted interest in a simple surgical procedure for sterilizing males (by severing the vas deferens duct) known as vasectomy.

Until the turn of the century, far more drastic castration was the only way to sterilize males. While castration was sometimes employed for therapeutic reasons, it was not popular with surgeons (one of whom was murdered by an irate patient) (Reilly 1991, 31). In 1899, a Chicago surgeon, A. J. Ochsner, described vasectomy in a paper appearing in the *Journal of the American Medical Association.* "Surgical Treatment of Habitual Criminals" reported improvements (including sexual function) in two patients on whom the procedure had been used to alleviate prostate problems and urged that it be extended to criminals. According to Ochsner, "There are undoubtedly born criminals." But normal individuals reared in criminal environments will also tend to develop lawless tendencies. Thus, whether criminals are born or made, it is desirable that they produce as few offspring as possible. Employing the procedure on criminals—as well as "chronic inebriates, imbeciles, perverts, and paupers"—would protect the community without harming the individual (Ochsner 1899, 867–68).

Dr. Harry Sharp of the Indiana State Reformatory soon expanded on Ochsner's suggestion. On the basis of vasectomies he had performed on 42 patients, Sharp lauded the procedure as a means to suppress masturbation (then thought to cause insanity and other ills) and proposed its use to counter the rapid increase in the number of criminals, paupers, insane, and feebleminded. Only in the human species, he argued, does the numbers of dependents rise, "since practically all other animal kind protect themselves, or are protected, by putting to death weaklings that are unable to weather the storm, and others that appear peculiar are castrated by their sires, and none but the perfectly healthy are left to reproduce their kind." Sharp urged legislatures to enact laws that would enable heads of all state institutions—including almshouses, insane asylums, institutes for the feebleminded, reformatories, and prisons— to "render every male sterile who passes its portals" (1902, 412–14).

Sharp himself did not wait. Between 1899 and 1907, and with no legal warrant, he sterilized nearly 500 males. (Males were targeted for a number of reasons: they filled the jails; it was assumed that many of the females were sterile as the result of venereal disease; the procedure for sterilizing females was relatively costly and dangerous [Reilly 1991, 34].) In 1907, Sharp helped persuade the Indiana legislature to enact the first statute authorizing compulsory sterilization of confirmed "criminals, idiots, rapists, and imbeciles." By 1912, sterilization laws had been enacted in eight states. The practice was defended on both humanitarian and financial grounds. Its proponents argued that it would allow many inmates to be discharged and, for this reason, save the taxpayers money. (It is notable that only one state statute applied to the noninstitutionalized.) The motivations of the American Eugenics Society are obvious from the section on segregation and sterilization in its 1926 pamphlet "A Eugenics Catechism":

Q. How much does segregation cost?
A. It has been estimated that to have segregated the original "Jukes" for life would have cost the State of New York about $25,000.
Q. Is that a real saving?
A. Yes. It has been estimated that the State of New York, up to 1916 spent over $2,000,000 on the descendants of these people.
Q. How much would it have cost to sterilize the original Jukes pair?
A. Less than $150.

Financial considerations were clearly paramount.

However, some eugenicists were unenthusiastic about sterilization, preferring segregation instead. Many people initially confused vasectomy with castration and thus associated sterilization with mutilation; Catholics were firmly opposed. Eugenicists thus feared that sterilization would give the movement an image of extremism and would risk jeopardizing its broader goals (Reilly 1991, 118–22). Moreover, they worried that vasectomies would increase levels of sexual activity, a situation undesirable in its own right. Popenoe and Johnson noted the opposition to sterilization on the grounds that it removed all fear of consequences and was thus "likely to lead to the spread of sexual immorality and venereal disease." They themselves maintained that the procedure was inhumane. "For society to sterilize the feeble-minded, the insane, the alcoholic, the born criminals, the epileptic, and then turn them out to shift for themselves, saying, 'We have no further concern with you, now that we know you will leave no children behind you,' is unwise. People of this sort should be humanely isolated. . . . Such a course is, in most cases, the only one worthy of a Christian nation; and it is obvious that if such a course is followed, the sexes can be effectively separated without difficulty, and any sterilization operation will be unnecessary" (1918, 185). Charles Davenport argued that sterilization laws were based on outdated scientific ideas and a category—feeble-mindedness—that was ill defined. Segregating defectives would be much more responsible (1911, 255–59). Others saw sterilization as complementary to segregation, the former to be used if the latter failed (Laughlin 1914, 45–47).

Legal challenges reduced the number of sterilizations more effectively than scientists' doubts did. Between 1913 and 1918, seven (of 12) state laws were constitutionally challenged; all of the challenges succeeded. Only a few operations were actually performed, mostly on the insane, who were much more likely than the retarded to be discharged from asylums (Reilly 1991, 48). Not until the 1927 case of *Buck v. Bell*, which upheld the Virginia statute, was eugenical sterilization practiced extensively. In his famous decision in that case, Justice Oliver Wendell Holmes wrote: "It is better for all the world if, instead of waiting to execute degenerate offspring for crime, or to let them starve for their imbecility, society can prevent those who are manifestly unfit from continuing their kind. . . . Three generations of imbeciles is enough." In the three years following the Supreme Court decision, 12 states enacted new laws or revised existing statutes. All had procedural safeguards that insured they would withstand constitutional challenge. Eventually, although 30 states adopted sterilization laws, more than half of the 60,000 individuals legally sterilized were in California.

During the Depression, budgets for custodial institutions shrank while the number of sterilizations grew. The public now found the procedure much more acceptable. A 1937 *Fortune* magazine survey of its readers found that 66 percent favored compulsory sterilization of mental defectives, 63 percent sterilization of criminals. Only 15 percent were opposed to sterilization of either (Reilly 1991, 125). The Popenoe and Johnson text reflects the changing attitudes that resulted from economic pressures. In 1918, the authors asserted that sterilization was an unsatisfactory substitute for segregation. In "states too poor or niggardly to care adequately for their defectives and delinquents" there might be no other choice, they conceded, but eugenicists "should favor segregation as the main policy" (Popenoe and Johnson 1918, 195). Their book was first revised in 1933, by which time Popenoe was secretary of the Human Betterment Foundation, the chief force behind eugenic sterilization in California. Although the authors claimed that they made no major changes in their book, they in fact wholly reversed themselves on the question of sterilization. In 1918, they had devoted less than seven (very critical) pages to the subject. In their revised edition, they accorded sterilization a very enthusiastic chapter of its own. It characterized critics of sterilization as ignorant, and the problems they identified as "imaginary." Segregation was now said to entail a "heavy expense" (Popenoe and Johnson 1933, 143, 152, 159).

Pressures on the public purse also explain the steady rise in the percentage of females who were sterilized; in a few states, *only* young women were sterilized—even though tubal ligation (the tying of the Fallopian tubes in order to prevent conception) constituted major abdominal surgery (Reilly 1991, 98). Economic pressures also explain the rush of sterilization laws in other countries.

By this time, of course, geneticists well recognized that preventing afflicted persons from breeding would not dramatically reduce the incidence of unwanted traits. "Where the incidence of a recessive defect in the general population is low to begin with (of the order of one-tenth of one per cent), even sterilization of *all* affected individuals in every generation reduces the incidence at a very slow rate, so that the eugenic value of the procedure may be questionable," wrote the geneticist Paul David. But that consideration did not lead him—or many others—to oppose sterilization. For he considered that "wholly aside from what germ-plasm the feeble-minded or recovered insane may or may not transmit to their offspring, there is the question as to whether these classes are fitted to discharge the rather intricate responsibilities of parenthood" (1933, 121). That line of reasoning was echoed by H. R. Hunt. Even if the mentally defective and insane have not inherited their defects, he argued, they are not fit "to rear a family anyway, so sterilization of the mentally ill and deficient is justifiable both on social and biological grounds" (1933, 151). Summarizing four decades of survey data (beginning in the early 1920s) and the comments of officials at various institutions, Philip Reilly concluded that during the Depression, state officials "became less concerned with preventing the birth of children with genetic defects and more concerned with preventing parenthood in those individuals who were thought to be unable to care for children" (1991, 94).

Eugenics in Germany and Scandinavia

In Germany, the American programs were frequently invoked as models. As early as 1913, a member of the Berlin eugenics society published a glowing report on the American eugenics movement. Géza von Hoffmann's *Racial Hygiene in the United States of North America* made many Germans (and Scandinavians, who closely followed German sources) envious of their American counterparts, who apparently enjoyed much greater popular and legislative support as well as success in attracting financial patrons (Kühl 1994, 15–18).

Before the Nazis came to power, German eugenics—like its American and European counterparts elsewhere—was both ethnically and politically diverse. The German Society for Race Hygiene, founded in 1905, was originally dominated by technocratic elitists alarmed by the declining fertility of the professional classes rather than by anti-Semites and ultra nationalists. Although the Society included racial purists, who formed a secret "Nordic Ring," they had to struggle against those who sought to purge eugenics of "unscientific" racism (Weiss 1987). A number of Jewish geneticists, such as Richard Goldschmidt, Franz Kallmann, and Curt Stern, were active eugenicists. As with eugenics movements elsewhere, so too in Germany; support came from political liberals and socialists as well as conservatives. But given its diverse supporters, the movement was inevitably riven with internal conflicts over both means and ends. What its members shared was a vision of the nation's health as a public resource and a belief that its problems could be solved through the breeding of a fit and healthy population or *Volk* (Weindling 1989).

The appeal of this technocratic vision had been greatly strengthened by the First World War. The nation's biological fitness now seemed crucial to its collective survival and thus a legitimate matter of state concern. In the war's devastating aftermath, that concern came increasingly to focus on cost-cutting. As elsewhere, hospitals and asylums were expensive to run and vulnerable to charges of excessive expenditures. In the context of these concerns, liberal and humanitarian values were increasingly swept aside (Weindling 1989).

In the aftermath of the First World War, close links were forged between German and American eugenicists. In the early 1920s, Harry Laughlin, superintendent of the Eugenics Record Office, began to report on German developments in the bulletin of the American Eugenics Society. In the late 1920s, he also began to report on American eugenics in German journals. The influential German geneticist Fritz Lenz established good relations with Laughlin, Charles Davenport, and Paul Popenoe (editor of the *Journal of Heredity* and coauthor with Roswell Johnson of the textbook *Applied Eugenics*). Popenoe in turn reported on American developments in the journal of the German eugenics movement (Kühl 1994, 18–20).

These relationships were not disturbed by the Nazi seizure of power. Popenoe continued to report favorably on German events. So did Laughlin, who used his position to organize the dissemination of Nazi propaganda. In 1935, the International Congress for Population Science met in Berlin. Two Americans—Laughlin and Clarence Campbell, president of the Eugenics Research Association—served as vice-presidents. (Marie Stopes, leader of the British birth control movement, also participated.) In their later reporting of events in Germany, they stressed that the law's intent was eugenic rather than racist, that it was carefully conceived, and that it included safeguards against abuse. Those themes were stressed by many other observers as well (Kühl 1994, 32–34, 44–52).

The Nazis regularly quoted American geneticists who expressed support for their sterilization policies. They also frequently invoked the large-scale California experience with sterilization. An analysis of that program by Ezra Gosney (the wealthy banker and citrus grower who founded the Human Betterment Foundation) and Paul Popenoe was published in 1929 as *Sterilization for Human Betterment*. Laughlin was a collaborator. The book lauded the California program as beneficial to those sterilized and cost-effective to the state. A German translation appeared the following year and was widely cited by leaders of the sterilization movement (Kühl 1994, 25, 43; Reilly 1991, 80–81, 106).

Before 1933, most German eugenicists had actually been dubious about proposals for compulsory sterilization, regarding them as politically unrealistic and scientifically premature (Weindling 1989). However, a draft law permitting sterilization with the consent of the person concerned or that person's guardian had been prepared in 1932, during the last days of the Weimar Republic. Before it could be approved, the government was in the hands of Adolf Hitler. In his manifesto, *Mein Kampf*, Hitler had proclaimed that the state "must declare unfit for propagation all who are in any way visibly sick or who have inherited a disease and can therefore pass it on, and put this into actual practice.... Those who are physically and mentally unhealthy and unworthy must not perpetuate their suffering in the body of their children" (1925, 404).

The Law for the Prevention of Genetically Diseased Progeny, issued two months after the Nazis came to power, allowed for compulsory sterilization, extended the range of "hereditarily determined" conditions, and required doctors to register cases of genetic disease (except in women past reproductive age). Sterilization was mandated, whether or not the person was institutionalized, in cases of congenital feeble-mindedness, schizophrenia, manic-depression, severe physical deformity, hereditary epilepsy, Huntington's chorea, hereditary blindness and deafness, and severe alcoholism. (Intelligence tests designed to determine feeblemindedness included such questions as "Why is there day and night?" and "Why does one build houses higher in towns than in the

FIGURE 5.2 "The Right Choice of Mate Is the Prerequisite for a Worthy and Prosperous Society." From brochure on race hygiene at the Hauptsarchiv Stuttgart; supplied by German historian Ute Deichmann.

countryside?") Genetic health courts were established to evaluate cases, which were usually referred to the courts by physicians. In 1935, the law was amended to allow for abortion within the first six months of pregnancy in the case of "hereditarily ill" women. Although not expressly permitted by the law, many ostensibly "asocial" persons were also sterilized. The genetic health courts found that deviation from the "healthy instincts of the Volk" constituted disguised or "social feeble-mindedness" and that sterilization of deviants was thus legal (Burleigh and Wippermann 1991, 136–97; Proctor 1988, 95–117). The Nazis also instituted a number of "positive" eugenics programs, such as loans and subsidies, to encourage breeding among favored groups. The best-known of these efforts was the "Well-of-Life" or *Lebensborn* program, which allowed single and married women who passed a racial test to give birth in special maternity homes run by the SS (Burleigh and Wipperman 1991, 65).

Sharing a common cultural ("Nordic") background with Germany, Scandinavia was in turn greatly influenced by German developments. Scandinavian eugenicists were bound to their German colleagues by strong personal, scientific, and institutional ties. Until 1933, their own movements developed along very similar lines. The ways in which they diverged after 1933 should alert us to the crucial importance of political context.

Eugenics was at least as popular in Denmark, Sweden, Norway, Finland, and Iceland as it was in Germany. In all the Scandinavian countries, sterilization laws were adopted almost without dissent. The enthusiasm for sterilization was in part a reflection of concerns about degeneration that Scandinavians shared with other Europeans and Americans. But that concern was not new in the late 1920s and the 1930s when the laws were adopted. What was new was the notion that management of the population might be technically and socially realistic (Frykman 1981, 48). In the perspective of the Social Democratic parties, eugenics seemed to go hand-in-hand with social reform (Roll-Hansen 1989). To see why, let us consider the case of Denmark, where the forces typically associated with the success of eugenics—racial and ethnic hostilities, social unrest, conservative opposition to working-class demands—were either absent or extremely weak (Hansen, in press).

The Social Reform Acts of the early 1930s marked the beginning of the Danish welfare state. Their chief architect, K. K. Steincke, was also instrumental in shaping Danish eugenics policy. In his 1920 book *The Welfare System of the Future*, Steincke questioned the value of improving social conditions in the absence of a program to control breeding. The provision of children's homes, kindergartens, sanatoriums, grants for child care, and other welfare programs, he argued, "weaken the stock by keeping thousands alive whom society had rather see succumb." By keeping defectives alive, "we give them the opportunity to bring tens of thousands of even poorer equipped and hereditarily burdened individuals into the world. . . . If we do not want modern European civilization to recede . . . we must systematically counteract the problematic consequences of civilization and also attempt to improve the race [by] the so-called eugenics, which is called racial hygiene or racial improvement in Scandinavia and Germany" (quoted in Koch 1993, 93).

Shortly after coming to power in 1924, the first Social Democratic government (with Steincke as justice minister) established a commission on castration and sterilization. Its report entitled "Social Measures against Degeneratively Predisposed Individuals" was published two years later. Wilhelm Johannsen, a distinguished geneticist who had also been a strong critic of the eugenics movement, served as a member of the commission. Realizing that the carriers of a gene for a rare recessive disease must vastly outnumber the afflicted, he explained that sterilization would have little eugenic effect. But like many Anglo-American geneticists who also understood this point, he did not thereby conclude that sterilization was wrong. The commission advised that "legislation directed towards a general racial improvement" was not yet feasible. But to sterilize institutionalized persons who were incapable of raising and educating children under acceptable conditions was nonetheless legitimate. Moreover, those children would themselves probably be genetically afflicted (Hansen, in press). As in the United States, Canada, and Britain, social and genetic arguments were intertwined. This same mix of genetic and social reasons was also characteristic of discussions in the other Scandinavian countries. Thus Swedish eugenicists stressed that sterilizing mental, physical, and moral defectives was desirable from society's point of view, even when the defect was not caused by heredity (Broberg and Tydén, in press).

In 1926, Denmark's Social Democratic government was replaced by a conservative one. Three years later, Denmark became the first Scandinavian country to pass a sterilization law, with only a very small group within the conservative party opposed. While in theory the law allowed only for "voluntary" sterilization, in practice this restriction was probably not very meaningful. In any case, the provisions regarding mental defectives were soon superseded by a 1934 law proposed by Steincke, now minister of health and welfare in the returned Social Democratic government. According to the new law, all mental defectives were to be confined in institutions, where they could be sterilized if judged unable to raise and support children or if the operation would facilitate their release from confinement. Application for sterilization would be made by a doctor and approved by an appointed guardian (most often the medical director of the custodial institution). The application would then be forwarded to a government department and, if approved, placed before a three-member board. Both in parliamentary discussions and in popular literature, the cost of maintaining defectives was a dominant issue (Hansen, in press).

While Danish eugenicists were sharply critical of Nazi racism and anti-Semitism, their praise for the German sterilization law was nearly unanimous. Like their counterparts in the United States and Canada, the Danes stressed the continuity with Weimar policy, the many formal safeguards in the new law, and the fact that indications for sterilization were wholly eugenic, rather than racial. Given the nature of their own policy, Danes were naturally disinclined to criticize the coercive aspect of the German law (Hansen, in press). Indeed, the German law did not appear extreme. Of course, its application was another matter. The most recent estimates of the number of Germans sterilized under the Nazi regime range from 320,000 to 400,000, with most sterilizations occurring during the first four years that the law was in effect and with little regard for legal niceties (Bock 1986, 230–46). In Denmark, by contrast, about 3,800 persons were sterilized between 1929 and 1945. Even adjusting for the smaller population (Denmark had one-twentieth of the population of Germany), the proportion of sterilizations in Denmark was vastly less. These facts show both that disparate social circumstances may produce similar laws and that similar laws may be applied very differently.

In Germany, the sterilization law proved to be a first step on the road to murder. But the experience of other countries shows that this is no necessary progression. In fact, the German program for sterilization virtually ground to a halt after 1939 while the number of sterilizations in Scandinavia was still on the rise. (The peak year for eugenical sterilization in Sweden was 1945 [Broberg and Tydén, in press]; the trend for Norway was similar [Roll-Hansen, in press].) Even today, all the Scandinavian countries still have sterilization laws little changed from those introduced in the 1930s. But these laws were never linked to a broader program of racial discrimination and extermination. The Scandinavians did not embark on programs to kill their mental patients. They did not forbid their Nordic citizens to marry those of "inferior" blood. They did not murder Jews, Gypsies, Slavs, and "asocial" persons.

In Germany, a registry of Gypsies was compiled by the psychiatrist Robert Ritter, chief of the Racial Hygiene and Population Biology Research Section in the Reich Health Department in Berlin. Ritter conducted research on their genealogy, concluding that Gypsies were an underclass of low hereditary value and a huge financial drain on the state (Müller-Hill 1988, 57). The register was used to round up Gypsies for transport to concentration camps, where those who did not die of starvation or disease were gassed. Lene Koch notes that the Danes, following Ritter, also produced a register of Gypsies, whom they, like the Germans, considered a social burden. But the Danes proposed to solve "the Gypsy problem" through their further integration with Danish society. In these two different social contexts, similar science was linked to opposite policies (Koch 1994, 2–3).

The German sterilization program was followed in 1939 by a euthanasia program designed to rid the nation of its mental patients, now characterized as "useless eaters." (The technology of the gas chamber was first developed in connection with this program [Proctor 1988, 114–15, 177–79].) Racist measures had been imposed shortly after Hitler came to power. In April 1933, Jewish and half-Jewish state employees and civil servants were dismissed from their jobs. In 1935, the first of the Nuremberg laws, defining who was a Jew, stripping them of citizenship, and forbidding their marriage with "citizens of German or related blood," was adopted. Other laws and extralegal actions directed against Jews, Gypsies, the offspring of German mothers and black French soldiers, homosexuals, and other social and political "deviants" followed, culminating in the program of mass extermination known as the Holocaust.

What was the attitude of German geneticists as their science was invoked on behalf of ever more extreme measures of racial purification? After the war, nearly all would claim that they too had been victims and suffered greatly under the regime. They knew nothing of the mass murders of mental patients and Jews. Even if they had joined the party, none had been Nazis "at heart." They found Nazi racism abhorrent. There were no traces of anti-Semitism in their work. Some of their best friends were Jews (Müller-Hill 1988).

Recent historiography tells a different story. Most of Germany's leading geneticists—including those who prior to 1933 had criticized anti-Semitism—actively helped construct the racial state. They served on important commissions, provided opinions on individuals' racial ancestry, gave courses on genetics for SS doctors, participated in the drafting of racial laws. More than half of all academic biologists joined the Nazi Party, the highest membership rate of any professional group (Deichmann and Müller-Hill 1994). That so many joined the Party (and also the SS and SA) is not explained by pressure; in fact, there was remarkably little. For example, party members and nonmembers had equal success in obtaining grants for their research. It rather reflects their enthusiasm for a regime that finally gave biologists, and geneticists in particular, the support they thought was their due. Far from being repressed, genetics—which was considered to be of great ideological, military, and economic importance to the regime—flourished in the Third Reich (Deichmann and Müller-Hill 1994; Macrakis 1993). Basic research was generously funded, career chances were expanded, and restrictions on experimental work were minimized (Deichmann 1992).

Free Love and Birth Control

The Nazi extermination programs stand at one extreme in the range of eugenic "solutions" to social "problems." The movements to liberate women from marriage stand at the other. In the late nineteenth century, many British social radicals such as Havelock Ellis, Olive Schreiner, and George Bernard Shaw promoted "free love" on the grounds that marriage, because it was so often based on social class or other nongenetic factors, was antieugenic. Shaw wrote the play *Man and Superman* (1903) to illustrate the point that "what we must fight for is freedom to breed the race without being hampered by the mass of irrelevant conditions implied in the institution of marriage." In Shaw's perspective, "What we need is freedom for people who have never seen each other before and never intend to see one another again to produce children under certain definite public conditions, without loss of honour" (1905, 74–75). In America as well, advocacy of eugenics and free love converged. Victoria Clafin Woodhull, the first American woman stockbroker and Equal Rights Party candidate for president in 1872, asserted the right of women to love as they pleased. Moses Harman, the revolutionary anarchist editor of the *American Journal of Eugenics*, also urged women to throw off the shackles of marriage. (When Harman died in 1910, unexpired subscriptions to his journal were completed with Emma Goldman's anarchist magazine, *Mother Earth*, which also distributed his books on eugenics.) In Denmark also, free love and eugenics were linked (Hansen, in press).

Arguments for sexual emancipation were often embedded in an evolutionary framework. Harman's daughter Lillian justified divorce on the grounds that the "state has barred the way of evolution . . . by holding together the mismated" (quoted in Gordon 1990, 120). Woodhull deplored the thwarting of sexual selection in contemporary society. She noted that "money bags are highly valued in the marriage mart," which accounted for the disturbingly large number of marriages between old men and young girls and between young men and old women. "When these marriages are fruitful," she explained, "they too often produce idiots, murderers, or otherwise unfit" (Martin 1891, 21). As we saw in chapter 2, Darwin's theory of sexual selection could be easily turned to a condemnation of capitalism or of economic inequalities between the sexes. In his novel *Looking Backward*, Edward Bellamy had argued that, in a classless society, gold would no longer "gild the straitened forehead of the fool." Instead of marrying the wealthiest men, women would choose the bravest, kindest, and most generous and talented, thus assuring the transmission of these desirable traits to posterity. Although the socialist Alfred Russel Wallace dismissed the importance of sexual selection in respect to other animals, he followed Bellamy in arguing that it would continuously raise the standard of the human race—if full equality of opportunity in respect to education and economic position were established (1890, 56; 1913). In 1898, the feminist socialist Charlotte Perkins Gilman proposed a variant of this argument claiming that, in the higher animal species, only women failed to choose their mates freely. Replacing the current social system with relations of economic equality was a prerequisite for the exercise of choice (Russett 1989, 84–86). That argument was echoed by Havelock Ellis, who claimed that in contemporary societies it is the woman who competes for the favor of the man. Women's economic independence, he wrote, "will certainly tend to restore to sexual selection its due weight in human development" (1914, 60). For these and many other socialists, economic equality was seen as a precondition for race improvement. Once achieved, hereditary progress would automatically follow (Gordon 1990, 126).

Thus sexual selection was often invoked as an alternative to negative eugenics. Shaw and Wallace both criticized proposals for state control of reproduction. "The world does not want the eugenicist to set it straight," asserted Wallace in an interview. "Give the people good conditions, improve the environment, and all will tend towards the highest type. Eugenics is simply the meddlesome interference of an arrogant, scientific priestcraft" (Wallace 1912, 177). Although Shaw argued that nothing but a eugenic religion could save civilization, when it came to practical policy he was wholly in agreement with Wallace (Shaw 1905, 74–75). Shaw did not trust political authorities to decide who should either be prevented from breeding or encouraged to breed. We should establish a socialist society, he argued, so that money and social position no longer determine who marries whom, and then "let people choose their mates for themselves, and trust to Nature to produce a good result" (Shaw 1928, 54). Virginia Woodhull, on the other hand, advocated negative eugenics as well as free love. In "Stirpiculture: or, The Scientific Propagation of the Human Race" (1888) and "The Rapid Multiplication of the Unfit" (1891), she argued that "the thoughtless, improvident, degenerate and diseased multiply upon us" and proposed that they be prevented from breeding (Martin 1891, 17).

In this respect at least, Woodhull's views converged with mainstream advocates of birth control, such as Margaret Sanger, doyenne of the movement in America, and Marie Stopes, her counterpart in Britain. In the late nineteenth century, it had often been claimed that voluntary motherhood would improve the quality of the human stock. Given the prevalence of Lamarckian assumptions about the inheritance of acquired characters, many reasoned that unwanted children were likely to be morally and/or physically defective. Indeed, the historian Linda Gordon notes that it would be hard to find a single piece of literature on voluntary motherhood between 1890 and 1910 that did not assert this claim (Gordon 1990, 119).

The argument that unwanted pregnancies produced defective offspring was not, in itself, racist or class biased. However, as concern about the differential birthrate mounted, claims about the quality of the biological stock took on a different and more menacing character. Birth control advocates came to stress the value of contraception in limiting births among the "unfit." As a consequence, the once politically radical movement gathered influential adherents but lost its working-class base.

In America, birth control advocates like Sanger and Emma Goldman had appealed to poor and powerless women with arguments that stressed their right to control their bodies, as well as the value of contraception in reducing maternal and infant mortality, the abortion rate, and the physical and economic burden of caring for children in the wretched circumstances of poor women's lives. Some also argued that large families weakened the poor in their struggle with capitalists. All demanded radical social and economic reform (Gordon 1990, 205). Sanger herself had joined the Socialist Party in 1911 but shortly left it in favor of the more radical Industrial Workers of the World. In a 1914 article in her paper *The Woman Rebel*, she defended an effort to blow up the Rockefeller family estate in Tarrytown, New York, in retaliation for the killing of striking miners and their wives and children in Ludlow, Colorado (Chesler 1992, 101–2). Indicted on charges of obscenity and incitement to murder and assassination, Sanger fled to Europe.

In Britain, she was strongly influenced by the Fabian socialists, especially Havelock Ellis. She returned to the United States a year later with some of the Fabians' positive attitudes toward experts and ambiguous attitudes toward workers. She began to advance the birth control cause by appealing to doctors, arguing that only they should have the authority to dispense contraceptive information. She tried to attract wealthy donors. And her writings began to vacillate between images of the poor as exploited and as unfit. She published heart-wrenching descriptions of the social conditions endured by the poor; their poverty, overwork, and fatigue are movingly portrayed. At the same time—sometimes on the same page—she characterized the poor as degenerates who represented an intolerable financial burden (Sanger 1922, 48).

Her British counterpart Marie Stopes, who had always been politically conservative, was less ambivalent toward the poor. As a member of the 1919 National Birth-Rate Commission, Stopes testified that the simplest way to deal with chronic cases of disease, drunkenness, and bad character was to sterilize the parents (Rose 1992, 134). On a 1921 visit to America, she complained of the millions of dollars spent every year on "armies of nurses, armies of doctors, armies of educators, armies of jail keepers, armies of reformatory masters, armies of protectors of the feeble-minded," when so little was done to stem the intolerable "stream of misery at its source" (Stopes 1921b, 9). In Stopes's view, it was disgraceful that middle-class taxpayers were able to rear only one or two children while society allowed "the diseased, the racially negligent, the careless, the feeble-minded, the very lowest and worst members of the community to produce innumerable tens of thousands of warped and inferior infants" (Stopes

1921a, 236). Just before the outbreak of war in 1939, Stopes expressed her admiration for Hitler by sending him a volume of her poems, but in 1942 she condemned the German enemy along with the Russians, Britain's Bolshevik allies. She wrote:

> Catholics, Prussians,
> The Jews and the Russians
> All are a curse
> Or something worse. . . . (quoted in Rose 1992, 219)

Unlike Sanger, whose attitude toward immigrants was sympathetic, Stopes was chauvinistic and racist.

In both Britain and America, birth control advocates tried to present their program as an alternative solution to the problem of the differential birthrate. As we have seen, numerous studies demonstrated a decline in middle-class fertility, with the steepest drop among female college graduates. Theodore Roosevelt and other opponents of contraception warned that the birthrates of college-educated women were already dangerously low and would decline further if birth control (or what was then often called "neo-Malthusianism") were easily available. In their perspective, contraception was difficult to encourage among the poor, while it was attractive and affordable to the middle class. A 1914 report of a committee of the American Breeders' Association expressed the typical view. Citing Roosevelt, the committee branded as "selfish" those of good stock who used artificial methods to prevent conception. It felt "constrained to condemn in no uncertain terms the purposeful limiting of offspring of parents of worthy hereditary qualities" (Laughlin 1914, 56).

Birth control advocates turned this argument on its head. They claimed that the middle class, unlike the poor and foreign-born, already had access to contraception. Eugenics thus depended on the spread of birth control information and methods. Reliable contraception, they promised, would reduce births among paupers, criminals, and other undesirables. By the early 1920s, most of the standard eugenic themes reverberate through Sanger's writings and those of contributors to the *Birth Control Review*, the magazine Sanger founded in 1917. (Its motto was "To create a race of thoroughbreds.") She now appealed to the authority of Karl Pearson, Charles Davenport, and Henry H. Goddard and to studies of the Jukes, the Kallikaks, and the army mental tests. She claimed that feeblemindedness—the "fertile parent of degeneracy, crime, and pauperism"—was rapidly increasing. The peril to future generations could only be averted if degenerates were prevented from reproducing their kind. "To meet this emergency is the immediate and peremptory duty of every State and of all communities" (1922, 81–82). Editorials and articles in the *Review* protested the "appalling rapidity" with which the unfit reproduced, and lamented their tremendous economic burden. According to Florence Tuttle, a frequent contributor to the magazine, "We have built asylums for the insane, institutions for the epileptic, and prisons even to punish them often for pre-natal sins, then applauded ourselves for having overcome evolution which would have refused to perpetuate this human wreckage. . . . The way to keep the social organism pure is to stop humanity's sewerage from emptying into it" (1921, 6). The tone of the movement is perhaps best reflected in the "Principles and Aims" of Sanger's American Birth Control League. They begin:

> The complex problems now confronting America as the result of the practice of reckless procreation are fast threatening to grow beyond human control. Everywhere we see poverty and large families going hand in hand. Those least fit to carry on the race are increasing most rapidly. People who cannot support their own offspring are encouraged by Church and State to produce large families. Many of the children thus begotten are diseased or feeble-minded; many become criminals. The burden of supporting these unwanted types has to be borne by the healthy elements of the nation. Funds that should be used to raise the standard of our civilization are diverted to the maintenance of those who should never have been born. (Sanger 1922, 279)

Sanger's rhetoric did not lead many eugenicists to support birth control. On the whole, they continued to believe that wider access to contraception would make a bad problem worse. But in respect to the wider public, the strategy was a success. The membership and budget of the American Birth Control League continued to grow. By the mid-1930s, contraception was essentially legal in the United States. Birth control was now respectable. It had also been converted, by means of eugenics, from a movement of social disruption to one of social control (Kennedy 1970, 121).

"Whose Country Is This?"
Eugenics and Race

I N FEBRUARY 1921, an article by Vice-President-elect Calvin Coolidge
entitled "Whose Country Is This?" appeared in *Good Housekeeping*
magazine. It addressed an issue then preoccupying Americans: should
immigration be limited and made more selective? Statutes to restrict
entry to foreigners had been proposed with increasing frequency during
the prior two decades. A bill under discussion in Congress (and adopted
three months later) would be the first to impose a numerical quota: 3
percent of the number of foreign-born of a given nationality as recorded
in the 1910 census. Coolidge rejected the view that America was
overcrowded, arguing that the country needed all the intelligence, skill,
and energy it could get. "There can not be too many inhabitants of
the right kind," he asserted. But lately, too many of the wrong kind
had been admitted. Indeed, America had become a "dumping ground"
(Coolidge 1921, 13–14).

Three years later, as president, Coolidge signed the Immigration Restriction (or "Johnson-Reed") Act into law. That statute limited the
number of allowable entrants to 165,000 annually—about a fifth of
the average prewar level. (In 1927, the number would be reduced to
150,000.) It also restricted entrants from any European country to 2
percent of the foreign-born of the same national origin recorded in
the 1890 census. Between 1880 and 1924, about 27,000,000 white
newcomers had arrived, some two-thirds of them from Russia, eastern
Europe, the Balkans, and Italy (compared with about 10,000,000
immigrants in the preceding sixty years, mostly from northwestern
Europe). Since most immigrants from southern and eastern Europe
arrived after 1890, the new system drastically reduced access to entrants
from that region (Shenton 1900, 265). (Canadians, fearing the thwarted
would now come to them, established categories of "preferred" northern

97

and "non-preferred" southeastern European immigrants [McLaren 1990, 58].) The House version of the bill had passed by a vote of 323 to 71; the vote on the Senate version was not recorded, but the bill had been approved by an overwhelming majority (Divine 1957, 185–92). The closing of the Open Door was an immensely popular act.

It had already been shut to Asians by a number of prior statutes. The Chinese Exclusion Act of 1882 had suspended the entry of laborers (with "laborers" defined to include all women) and prohibited all foreign-born Chinese from acquiring citizenship. (The suspension was made "permanent" in 1902.) The provision barring citizenship was based on the Nationality Act of 1790, which limited naturalization to "free white persons," a law amended in 1870 to include aliens of African birth and descent. From the congressional debates in 1870, which focused on the menace of Chinese immigration, and the overwhelming defeat of a proposed amendment to extend naturalization laws to "persons born in the Chinese empire," it is clear that Asians were intentionally excluded (Hutchinson 1981, 57–58).

In 1907, the "Gentleman's Agreement" with Japan called for the "voluntary" restriction of immigration in return for an end to the segregation of Japanese pupils in San Francisco schools. And in 1917, an Asiatic Barred Zone was created, excluding laborers from the rest of Asia. In a largely gratuitous slap at the Chinese and Japanese, the Johnson-Reed Act barred all aliens who were ineligible for citizenship. It thus reaffirmed Chinese exclusion and abrogated the Gentleman's Agreement, halting the small trickle of Japanese emigrants (who had been made racially ineligible for citizenship by a Supreme Court decision in 1922) (Bernard 1980, 493).

The Chinese Exclusion Act was the first federal statute to bar entry to a group specifically for reasons of race. About 300,000 Chinese had arrived on the West Coast in the 1850s, 1860s, and 1870s, mostly as contract laborers to work on the railroads. The initial welcome soon changed to antipathy, especially on the part of organized labor. Throughout the 1860s, both the Democratic and Republican party platforms had regularly praised foreign immigrants (who had "added so much to the wealth, development of resources, and increase of power to the nation," according to the Republican platform of 1864) and endorsed liberal policies of immigration and naturalization. But by 1876, both parties sounded a different note concerning the Chinese. The Democratic platform blamed the government for having "exposed our brethren of the Pacific Coast to the incursions of a race not sprung from the same great parent stock," people who were "neither accustomed to the traditions of a progressive civilization nor exercised in liberty under equal laws" (Hutchinson 1981, 624). The fear was twofold: that the Chinese would threaten standards of wages and working conditions and that they would be racially unassimilable. But the act barring them from entry had even more far-reaching consequences, for it established the right of the federal government to restrict the immigration of particular national groups (Bernard 1980, 490). It set the ideological stage for the legislation Coolidge signed in 1924.

Disparate economic, political, cultural, and biological anxieties converged to produce the clamor for restriction. Workers feared economic competition, while their employers feared industrial unrest. Many native-born Americans of every social class and political persuasion abhorred the aliens' customs and suspected their political loyalties. The native-born also blamed the newcomers for increasing levels of crime, insanity, and pauperism, and thus for the financial burden of custodial care, for urban political corruption, for strikes and other forms of labor militancy, and for unemployment, among other social ills. The new immigrants were thought to be culturally and biologically ill adapted to life in America—and rapidly breeding. The various (and sometimes contradictory) strands in this indictment are not always easy to separate.

Resentment of economic competition from aliens was exacerbated by the severe unemployment that plagued the country following the end of the First World War. The American Federation of Labor was especially anxious about the effects of competition on employment, working conditions, and wages. (The conservative Coolidge claimed that there were plenty of jobs available and blamed strikes for the fact that so many laborers were out of work.) Even a number of socialists favored restriction on economic grounds; they claimed that capitalists induced immigration to undercut unions and that southern European and Asian immigrants (as well as blacks) were impossible to organize (Pittinger 1993, 169). In one of the more blatant examples of the adage about politics making strange bedfellows, socialists were de facto allies of capitalists who hoped to curb labor unrest.

The First World War also promoted a heightened sense of nationalism, with "bad" Europe counterpoised to "good" America (Daniels 1990, 278). The loyalty of immigrants came under suspicion. The war itself was followed by the Red Scare, a period of hysteria over "enemies within." While steamship company owners and industrialists opposed immigration restriction, which would cut off the supply of passengers and cheap labor, other businessmen, noting the high proportion of new immigrants in radical labor organizations, supported such legislation as a cure for industrial and political militancy. Coolidge reflected the common tendency to associate immigrants with anarchism and Bolshevism when he asserted that "there is no room for the alien who turns toward America with the avowed intention of opposing government, with a set desire to teach destruction of government—which means not only enmity toward organized society, but toward every form of religion and so basic an institution as the home" (Coolidge 1921, 14).

Cultural issues were at least as salient as economic and political issues. Language, religion (most of the European newcomers were Catholic or Jewish), and social customs separated the new immigrants from the native-born. Across the political spectrum, the newcomers were viewed as indigestible lumps in the Melting Pot. Thus union members who had little in common with Calvin Coolidge could endorse his view that American institutions flourished only by reason of common experiences and values that were not shared by the new immigrants, that "there are many who land here who really never get to America" (Coolidge 1921, 106).

Moreover, the immigrants appeared to have very large families. The 1910s and 1920s witnessed an outpouring of scholarly and popular works lamenting the high fertility of immigrant groups, especially in comparison with the birthrates of native-born Americans. Numerous geneticists joined the chorus, providing scientific credence to the cries of "race suicide." The geneticists often focused on the relative reproductive rates of immigrant groups and graduates of elite colleges (such as themselves). The distinguished authors of one textbook are typical of writers in this genre. They wrote:

> A Harvard class does not reproduce itself and at the present rate, one thousand graduates of today will have only fifty descendants two hundred years hence. On the other hand, recent immigrants and the less effective descendants of the earlier immigrants still continue to have large families; so that from one thousand Roumanians today in Boston, at the present rate of breeding, will come a hundred thousand two hundred years hence to govern the fifty descendants of Harvard's sons (Castle et al. 1912, 309).

"Race Suicide"

No academic did more to popularize fears of "race suicide" than Edward A. Ross, the reform-minded sociologist who coined the term (Ross 1901). In his influential book *The Old World in the New*, Ross warned that "low-standard immigrants"—especially Russians, Hungarians, Poles, Southern Slavs, Italians, Greeks, and Portuguese—were reproducing much faster than more valuable, older-stock Americans. "The fewer brains

Published Weekly

The Curtis Publishing Company

Cyrus H. K. Curtis, President
C. H. Ludington, Vice-President and Treasurer
F. S. Collins, General Business Manager
Walter D. Fuller, Secretary
William Boyd, Advertising Director

Independence Square, Philadelphia

London: 6, Henrietta Street
Covent Garden, W.C.

THE SATURDAY EVENING POST

Founded A°D¹ 1728 *by* Benj. Franklin

Copyright, 1921, by The Curtis Publishing Company in the United States and Great Britain
Title Registered in U. S. Patent Office and in Foreign Countries

George Horace Lorimer
EDITOR

Churchill Williams, F. S. Bigelow,
A. W. Neall, Arthur McKeogh,
T. B. Costain.
Associate Editors

Entered as Second-Class Matter, November 19,
1879, at the Post Office at Philadelphia,
Under the Act of March 3, 1879
Additional Entry as Second-Class Matter
at Columbus, Ohio, and at Decatur, Illinois
Entered as Second-Class Matter at the
Post-Office Department, Ottawa, Canada

Volume 193 | 5c. THE COPY 10c. in Canada | PHILADELPHIA, PA., APRIL 30, 1921 | $2.50 THE YEAR by Subscription | Number 44

The Existence of an Emergency

By Kenneth L. Roberts

Testimony taken by the Senate Immigration Committee in hearings on the Johnson Bill prohibiting immigration for one year has failed to prove the existence of an emergency, according to senators who analyzed evidence submitted by more than thirty witnesses. One member said that the bill would be sidetracked until the emergency could be proved.—*Cable dispatch, January 14, 1921.*

EMERGENCIES are frequently determined by the point of view. A small American city which wakes up one morning to find itself with two hundred cases of scarlet fever and fifty cases of diphtheria on its hands is very apt—in view of the emergency—to throw a series of epileptic fits, froth largely at the mouth with fright and disinfect everything from the ornamental weather vane on the church to the pyramidal pile of iron cannon balls on the front lawn of the G. A. R. Hall. By these and other protective measures the epidemic is arrested and the emergency dies a sudden and enthusiastic death.

Around the middle of January, 1921, I had occasion to investigate the state of affairs in a Polish town which had a population of 8000. There were a number of cases of typhus in the town. The residents themselves didn't know how many, but they thought the number would be about three hundred. That number, according to them, wasn't worth considering. Typhus, like the poor, they had always with them. There was a great deal of diphtheria and scarlet fever in the town. The residents couldn't tell me exactly how much; but there were several hundred cases. Nothing serious, they insisted; only a few hundred.

Less than three miles away there was another town, they explained, where things were worse. Practically every family in that town had at least one case of typhus in it. Up to the north and northeast there was cholera; but it hadn't reached town—yet. No emergency existed for the people of that town, in their judgment. For them typhus and scarlet fever and diphtheria were old stories. Each year a certain number of people had to have them and a certain number had to die of them, just as part of the potato crop had to freeze in the ground. Sickness was a part of the scheme of things. The person who suggested to them that they were confronted by an emergency would have been responsible for a number of cracked lips. As a matter of fact, a perpetual emergency existed for them. In the same way the United States has been confronted by an immigration emergency for years. Starting around 1880 the immigrants who swarmed into the United States were of an entirely different breed from the people who had discovered the country, colonized it, made its laws and developed it. These new and different people came in waves, like the waves of an endlessly rising tide. Occasional waves fell short of preceding waves; but in general they surged to higher and higher levels. In the year 1905 more than a million of them came—more than forty army divisions. It was the same in 1906. In 1907 the wave surged over the million-and-a-quarter mark. In the ten years which ended June 30, 1914, more than ten millions of these people had entered the United States. They sank naturally into the slums and the foreign settlements; for the great percentage of them had always lived in either city or agricultural slums, and practically all of them had come to America to make as much

Herbert Johnson

Look Out for the Undertow!

FIGURE 6.1 Front Page of *The Saturday Evening Post,* April 30, 1921. Photo courtesy of the Boston Public Library. Kenneth Roberts frequently published diatribes against immigrants from southern and eastern Europe in the *Saturday Evening Post,* one of America's leading popular magazines. His 1922 book *Why Europe Leaves Home* described Polish Jews in particularly unflattering terms.

they have to contribute, the lower the place immigrants take among us," he asserted, "the faster they multiply" (1914, 299).

Ross was also intellectual mentor to Theodore Roosevelt, who shared his alarm at the breeding rates of "subnormal" types. Roosevelt thought it "obvious that if in the future racial qualities are to be improved, the improving must be wrought mainly by favoring the fecundity of the worthy type and frowning on the fecundity of the unworthy type" and complained that we did the reverse (Laughlin 1914, 56). Like Ross, he was less concerned by fecundity among the poor than among nations and immigrant groups he considered inferior. Roosevelt had been exposed to racialist theories as an undergraduate at Harvard and a law student at Columbia (Dyer 1980, 5–6). In later life, he read voraciously in the literature of race and consulted with a number of prominent eugenicists, including Charles Davenport. Given his enormous popularity, Roosevelt probably did more than any other individual to bring the views of academic race theorists to ordinary Americans.

In hundreds of articles and speeches, Roosevelt attributed America's success to its (originally) good Anglo-Saxon blood. He spoke constantly of the country's "race destiny," which he thought was threatened by both the influx of inferior immigrants from southern and eastern Europe and the failure of old-stock Americans to reproduce at an adequate rate. Americans needed to "keep out races which do not assimilate with our own," he asserted. During his presidency (1901–09), Roosevelt repeatedly called for curbs on immigration. At the same time, he argued that old-stock American families had a duty to produce four to six children (Dyer 1980, 14–15). As governor of New York, and later as vice-president and president, Roosevelt constantly preached that Americans of English blood, and Anglo-Saxons generally, were involved in a desperate "warfare of the cradle" with lesser races. If those of Anglo-Saxon stock were to prevail, it was essential that women "not flinch from breeding." Even those already physically depleted by multiple births should act like soldiers and place their reproductive duty first. In Roosevelt's view, both men and women of good stock who chose not to have children were "race criminals." He even proposed a constitutional amendment placing marriage and divorce law under federal authority in order better to protect the home life of the nation. In his 1906 message to Congress, Roosevelt warned that deliberate sterility was:

the one sin for which the penalty is national death, race death; a sin for which there is no atonement; a sin which is the more dreadful exactly in proportion as the men and women guilty thereof are in other respects, in character, and bodily and mental powers, those whom for the sake of the State it would be well to see the fathers and mothers of many healthy children. (Roosevelt 1926, 376–77)

As it turned out, the assumption of high immigrant fertility was wrong. While immigrants did generally produce more children than native-born women, their daughters did not. In fact, the fertility of these second-generation women, many of whom apparently delayed or avoided marriage in order to work and help support their families, was exceptionally low (King and Ruggles 1990, 347–69). But it seemed self-evident from the declining fertility of Yankee stock combined with the huge foreign presence in cities that immigrants were outbreeding native-born Americans. By 1920, the total population of the United States was 105 million; of these 14 million were foreign born, while another 22 million had at least one foreign-born parent. Immigrants and their children constituted more than a third of the entire population (Daniels 1990, 274–75). Thus old-stock Americans were indeed being swamped, especially if they resided in cities. By the turn of the century, over half of the residents of the northeastern states were of foreign parentage; in some cities, such as New York, over three-quarters were. Since most of these aliens were born in America, the notion that immigrants were exceedingly fertile seemed obvious (King and Ruggles 1990, 366).

As the preceding discussion suggests, cultural and biological anxieties were often intertwined and mutually reinforcing. In 1910, the congressionally appointed Dillingham Committee (named after its chair, Senator William P. Dillingham, a Vermont Republican) issued a 42-volume report on the effects of immigration. Its findings were alarmist. The commission produced a mass of statistics in support of its claim that the new immigrants were racially inferior to the old.

The bible of the restrictionist movement was Madison Grant's 1916 polemic *The Passing of the Great Race.* Grant was a wealthy and socially prominent lawyer (Yale class of 1889), chairman of the New York Zoological Society, member of many social clubs, and vice-president of the influential Immigration Restriction League, whose members were mostly elite, Harvard-educated Bostonians (Solomon 1956). In a letter recommending him for admission to the Century Club, the writer remarked that Grant's "strong views on certain questions, like Catholicism and the Hebrew race, have made him some enemies" (quoted in Chase 1977, 164). His book sang the praises of the "great race" of Nordics, a "virile" people characterized by tall stature, light colored eyes, fair skin, blond or light brown hair, straight noses, and "splendid fighting and moral qualities." It vilified the ever-increasing number of new immigrants (characterized as "moral perverts, mental defectives, and hereditary cripples" [Grant 1916, 45, 75, 81–82]).

The mixing of these groups would only result in racial decline. Grant lamented the "unfortunate fact that nearly all species of men interbreed." Only strongly marked racial feeling could prevent Nordics from mating with those of inferior stock. "Race feeling may be called prejudice by those whose careers are cramped by it," he wrote, "but it is a natural antipathy which serves to maintain the purity of the type." If not kept apart by such distinctions, races "ultimately amalgamate, and in the offspring the more generalized or lower type prevails" (1916, 193).

Grant attributed fertility declines among the native-born to their refusal to bear children who would have to compete with repulsive newcomers. The native American, "too proud" to mix with these aliens, was thus abandoning "the land which he conquered and developed." In an oft-cited passage, Grant wrote that old-stock Americans are "today being literally driven off the streets of New York City by the swarms of Polish Jews. These immigrants adopt the language of the native American; they wear his clothes; they steal his name; and they are beginning to take his women, but they seldom adopt his religion or understand his ideals, and while he is being elbowed out of his own home the American looks calmly abroad and urges on others the suicidal ethics which are exter-minating his own race." Grant blamed democracy for this debacle, for wherever it was adopted, the "genius of the small minority is dissipated" and power shifted from the Nordic to racial inferiors (1916, 8, 81).

Grant represents an extreme of the restriction movement, but his views were still considered respectable. Henry Fairfield Osborn, paleontologist, conservationist, President of the American Museum of Natural History, and Da Costa Professor of Zoology at Columbia, provided the preface to Grant's book, which was frequently cited and received generally favorable reviews in scientific journals. In a discussion in the prestigious journal *Science*, Frederick Adams Woods noted that no sources were provided for any of the book's claims but nonetheless pronounced it "a work of solid merit." He found Grant's theory of the special value of Nordic blood largely convincing (1918, 419). In a later article, he noted that most of the unfavorable reviews were "signed by persons of non-Nordic race" (1923, 95). By this time *The Passing of the Great Race* was in its fourth edition. In 1933 Earnest Hooton, America's leading physical anthropologist, refused Grant's request to review his new book, *The Conquest of the Continent*. Hooton replied: "I don't expect that I shall agree with you at every point, but you are probably aware that I have a basic sympathy for you in your opposition to the flooding of this country with alien scum" (quoted in Barkan 1992, 313). Charles

Davenport expressed even stronger views. In a letter to Grant written shortly after passage of the Johnson-Reed Act, he wrote: "Our ancestors drove Baptists from Massachusetts Bay into Rhode Island but we have no place to drive the Jews to. Also they burned the witches but it seems to be against the mores to burn any considerable part of our population. Meanwhile we have somewhat diminished the immigration of these people" (quoted in Chase 1977, 301).

Grant's influence was exerted as much through his personal contacts as his publications. As a member of the Boone and Crockett Club of New York, he socialized with its president, Theodore Roosevelt (Chase 1977, 163). Grant also corresponded with Roosevelt, who endorsed his views. John Merriam, the president of the Carnegie Institution of Washington (which funded the Eugenics Record Office under Davenport and Laughlin) was a close friend. So was Elihu Root, recipient of the 1912 Nobel Peace Prize, secretary of war and of state in Roosevelt's cabinet, U.S. Senator from New York, and one of the most influential trustees of the Carnegie Institution (Lagemann 1992, 80). Grant had close contacts with the physical anthropologist Aleš Hrdlicka, curator of Anthropology at the Smithsonian Institution and editor of the *American Journal of Physical Anthropology* (Barkan 1992, 97). As a member of the Eugenics Committee of the United States of America, a self-appointed group that developed into the American Eugenics Society, he chaired the Committee on Selective Immigration. Its members included Harry Laughlin (Charles Davenport's associate, and "superintendent" of the Eugenics Record Office at Cold Spring Harbor) and Congressman Albert Johnson, a Washington Republican who chaired the House Committee on Immigration and Naturalization. In 1921, Laughlin was appointed "Expert Eugenics Agent" to Johnson's committee.

Through his personal connections and writings, Grant legitimized the view that crossing of Nordic with new immigrant stock would produce biological degradation. Calvin Coolidge certainly echoed Grant when he wrote that "there are racial considerations too grave to be brushed aside for any sentimental reasons. Biological laws tell us that certain divergent people will not mix or blend. The Nordics propagate themselves successfully. With other races, the outcome shows deterioration on both sides" (1921, 14). In 1921, this view had become such a commonplace that a holder of high office could assert it without shocking his audience. Indeed, it was a commonplace among geneticists. Only one geneticist— H. S. Jennings—could be roused to oppose Laughlin's testimony publicly. How can we explain such a view taking root in a democratic country?

Evolution, Genetics, and Race

We saw in chapter 2 that Darwin and most of his peers assumed that natural selection explained physical, mental, and moral differences among races. Also, as we saw in chapter 3, biology and culture were not sharply distinguished in the nineteenth century. Especially in the Lamarckian view, where cultural practices would necessarily have biological consequences, no clear distinction was drawn between nations, "races" defined by skin color, and what today are called ethnic groups. A "race" was any population with common attributes that was also linked by descent. Nineteenth-century literature is replete with references to the French, Russian, Jewish, and Slavic "races"—describing peoples with a common language, religion, and history. "Anyone who considers the Jews will see at once that their character, as much as their noses, are an inheritance," according to one writer in 1885. "A Scotchman 'caught young' . . . may lose some of the superficial characteristics, but will retain all the national peculiarities of his race: and so will the Irishman" (Wheeler 1885, 499). "Race" was also employed more broadly, as in the "Anglo-Saxon" or human race. In the early twentieth century it sometimes defined a group linked by descent that differed from others in a single trait. Thus Charles Davenport wrote that "a blue-eyed Scotchman belongs to a different race from some of the dark Scotch" (Davenport 1917a, 364). The concept of "race hygiene" and phrases like "racial efficiency," and "racial vigor" were therefore highly ambiguous. So were the cries of "race suicide," "race death," and "race decadence," which could refer to either an internal or external threat.

Darwin certainly did not invent the view that nations differ in mental, moral, and temperamental qualities. Moreover, fears of racial degeneration preceded his work. But his explanation of national differences in terms of natural selection reinforced the view that societies could be scientifically ranked along a continuum ranging from the most barbaric to civilized. It seemed also to imply that higher populations would displace the lower. Darwin himself was disturbed by the dependence of progress on such a harsh competitive struggle. But even those who regretted the brutality of the process thought the extinction of native peoples ultimately for the best.

Given their assumptions about the outcome of competitive struggle, Darwinians did not fear being swamped by these peoples (as they did fear the domestic poor). Thus in Britain and other parts of Europe, evolutionary racism functioned to justify imperial authority and strenuous efforts in the colonies to prevent "contamination" of European by native blood. Given the small number of European women in the colonies, race-mixing was a constant threat (Stoler 1991, 75). But Britain, France, Holland, Belgium, Spain, and other European powers had relatively homogeneous populations. Eugenics at home thus focused on class, not race.

In the United States, with both its large internal black population and its influx of immigrants from Asia and from southern and eastern Europe, the situation was different. While evolutionary theory served to justify imperialist policies in the Philippines, Cuba, and Puerto Rico, it also sanctioned racist domestic policies, including segregation and antimiscegenation laws and efforts to restrict immigration (Bannister 1979). The Chinese (like blacks and native Americans) were characterized as peoples who had reached the end of the evolutionary road (Pittinger 1993, 174).

Of course the Chinese were also condemned on racial grounds by many individuals who would never have thought to base their arguments in evolutionary theory. President Grover Cleveland was unlikely to have been thinking of Darwin when he asserted in 1882 that the "experiment of blending the social habits and mutual race idiosyncrasies of the Chinese laboring classes with those of the great body of the people of the United States ... [has been] proved ... in every sense unwise, impolitic, and injurious to both nations" (quoted in Daniels 1990, 272). Evolutionary theory was only one element in a complicated tapestry of racist arguments.

Those who did invoke evolutionism in support of restrictionist views were of all political persuasions. But socialists, in their enthusiasm for science in general and Darwinian materialism in particular, were particularly attracted to evolutionary arguments. Evolutionary claims and images suffused debate on many issues of socialist tactics and strategy and were particularly marked in the debate over Asiatic exclusion (Pittinger 1993, 174–79).

Within the Socialist Party, such prominent figures as the novelist Jack London and the theorist Robert Hunter issued dire warnings of race suicide. Hunter predicted that unless the native birthrate increased and the influx of newcomers subsided, the immigrants' prolific breeding would result in "the annihilation of the native American stock" (quoted in Pittinger 1993, 173). Evolutionary arguments were invoked most frequently on the right wing of the Socialist movement, where Darwinism was used to argue for social evolution in preference to revolution. Some leftists, such as Eugene Debs, condemned racialism and immigration restriction. But their influence was never strong, and it waned over time. At the 1912 Socialist convention, the majority report characterized racism as "a product of biological evolution, a complex of feelings more deeply rooted than class and certain to persist under socialism." The report argued that racism arose as a protective barrier against amalgamation wherever races struggled to survive, and it saw race wars as normal mechanisms of natural selection. To oppose restriction was therefore utopian and unscientific (Pittinger 1993, 178).

Intelligence Testing

Darwinism was but one weapon in the arsenal of scientific arguments for immigration restriction. Mental testing was another. In 1913, Goddard began to administer a form of the Binet-Simon exam at Ellis Island. The results were dramatic. On the most favorable assumptions, it appeared that 40 percent of recent immigrants were feebleminded. Goddard himself did not presume that the cause was genetic; indeed, he considered it "far more probable that their condition is due to environment than that it is due to heredity" (Goddard 1917, 270). He noted that the immigrants' environments had been poor and that there had been no noticeable increase in the proportion of feebleminded of foreign ancestry; less than 5 percent of the inmates of institutions for the feebleminded were of foreign parentage. In any case, with recent immigrants the effects of heredity and environment would be hopelessly entangled. What really alarmed Goddard were the "degenerate" lines of rural Anglo-Saxon stock. The Kallikaks and the Jukes, as well as other families that had been studied, such as the Pineys, Dacks, Yaks, and the Smoky Pilgrims, were hillbilly families of old English stock, not urban immigrants (Rafter 1988). While Goddard's provisos were sometimes ignored, his 1917 article "Mental Tests and the Immigrant" had little direct impact on the restrictionist cause. Its importance lay elsewhere, in the expansion of mental testing. Goddard himself was involved in the development of the army mental tests, which appeared to indicate that 45 percent of foreign-born draftees had a mental age of less than eight (as compared with 21 percent of the native-born). To a modern reader, the cultural bias of the army tests is obvious. Draftees were required to identify authors, athletes, and actresses, fictional characters in books and advertisements, the names of card games, the sites of Civil War battles, and types of American automobile engines, among other items clearly dependent on education or experience. Nevertheless, the test results were cited as evidence that a stream of "defective germ plasm" was flooding into America from eastern Europe, Russia, the Balkans, and Italy.

In 1923, two books—Lothrop Stoddard's *The Revolt against Civilization* and Carl C. Brigham's *A Study of American Intelligence*—helped popularize the message. Stoddard was a Harvard graduate and prolific propagandist who warned in dramatic language of the threat from barbarians both within and outside the gates. In F. Scott Fitzgerald's *The Great Gatsby*, Tom Buchanan accurately reflects the spirit of Stoddard's 1920 *The Rising Tide of Color against White World-Supremacy* (which included an introduction by Madison Grant). "Civilization's going to pieces," Buchanan suddenly announces to his wife and their dinner guest. "I've gotten to be a terrible pessimist about things. Have you read 'The Rise of the Coloured Empires' by this man Goddard? . . . The idea is if we don't look out the white race will be—will be utterly submerged. It's all scientific stuff; it's been proved. . . . It's up to us who are the dominant race to watch out or these other races will have control of things" (Fitzgerald 1925, 17). In a 1921 speech in Birmingham, Alabama, President Warren Harding cited Stoddard in support of his determined rejection of "every suggestion of social equality" between blacks and whites (quoted in Tucker 1994, 93). In *The Revolt against Civilization*, Stoddard invoked the army tests as proof that prolific alien stocks were a deadly menace to American civilization. Intelligence is being "bred out of the race," he warned. If the decline is not stopped, civilization will "crash from sheer lack of brains" (Stoddard 1923, 114).

Stoddard was an amateur historian; Brigham, a professional psychologist. Academics who might have hesitated to associate themselves with the author of popular books written in sensational language had no compunctions about citing the sober and respected Brigham. *A Study of American Intelligence* included an introduction by Yerkes warning that "no citizen can afford to ignore the menace of race deterioration." Brigham assured his readers that it had been proven beyond doubt that "inferior peoples or inferior representatives of peoples" were flooding into America (Brigham 1923, 204–5). He explicitly endorsed Madison Grant's claims of Nordic superiority, and he lamented the mixing of Nordic with Slavic, "degenerated hybrid Mediterranean," and even Negro blood. According to all the available evidence, he asserted, "American intelligence is declining, and will proceed with an accelerating rate as the racial admixture becomes more and more extensive. The decline of American intelligence will be more rapid than the decline of the intelligence of European national groups, owing to the presence here of the negro. These are the plain, if somewhat ugly, facts that our study shows." Brigham concluded that national intelligence would continue to decline even if all immigration were stopped immediately. While immigration restriction was needed, "the really important steps are those looking toward the prevention of the continued propagation of defective strains in the present population" (1923, 182, 210).

Biologists and Race-Mixing

Mental tests were certainly a factor in the debate over immigration restriction, but their impact should not be exaggerated. The movement to restrict immigration had been gathering steam since the 1890s. Articles in popular magazines warning that the new groups were incompatible with the old only rarely cited test results. Moreover, tests played no role in the movement to restrict Asian immigration (Gelb 1985). Through Laughlin, the arguments of mental testers were called to the attention of congressional leaders such as Albert Johnson. But Johnson did not need to be convinced; he was already a Nordic supremacist when he appointed Laughlin adviser to his committee. Thus, in respect to passage of the Johnson-Reed Act in 1924, the activities of mental testers served mostly to reinforce a larger biological argument about the dangers of race-mixing. Coolidge, like Johnson, did not need test results to know that the crossing of some groups produces biological deterioration. That argument had been made long before the advent of mental tests. Within limits, it was sometimes claimed, the mating of those with different "customs, habits, feelings, thoughts" had positive effects. But the crossings must be judicious. "In the human species it is probable that the crossing of those varieties called national varieties, even strong national varieties, produces good results," wrote the American evolutionist Joseph LeConte in 1879, "but the crossing of varieties so divergent as those called primary races is probably bad" (LeConte 1879, 177–78). This argument was made more precise by the work of twentieth-century geneticists and physical anthropologists.

In the eighteenth and nineteenth centuries, plant and animal breeders knew that inbreeding (the mating of close relatives) was often accompanied by deterioration and, conversely, that crossing between closely related strains was accompanied by an increase in "vigor." In Darwin's day, breeders would say that parents from different strains usually possess different defects that tend to cancel out in their progeny. After 1900, that explanation was rephrased in Mendelian terms. Edward East, the world's leading authority on the effects of inbreeding and outbreeding, explained that "inbreeding may weaken a house by uncovering a whole closet full of family skeletons—recessive abnormalities" (East 1927, 15). Thus outbreeding was thought to be generally good for plants and animals.

But did this conclusion extend to humans? Like LeConte, most scientists thought so—within limits. To a certain degree, eugenics actually fostered a positive attitude toward race-mixing, as outcrossing increases the heritable variation on which selection depends. Thus Davenport wrote: "The person who seeks to secure the racial improvement of any species has only two courses open to him: either to await new mutations in the race which he wishes to improve, or else to cross it with some other race that already has the quality he desires" (Davenport 1928, 238; see also East 1927, 15). Mutation is slow, and the eugenicists were impatient. In theory, race-crossing would speed up eugenic selection.

Arguments for and against Miscegenation

East wrote glowingly of the benefits of race-crossing: "The great individuals of Europe, the leaders in thought, have come in greater numbers from peoples having very large amounts of ethnic mixture." The back-wardness of Spain and Ireland was ascribed to their relative isolation. East called for Jews to give up their "cult of racial purity" and attributed America's vigor to its Open Door policy (East and Jones 1919, 257, 261).

Nonetheless, he urged that the policy be rethought in light of the changed character of recent immigration. "The ingredients in the Melting Pot must be sound at the beginning," he wrote, "for one does not improve the amalgam by putting in dross" (East and Jones 1919, 269). East thus called for a pause while the worth of these newcomers was evaluated. He was certain there was no benefit from adding blacks to the "White Melting Pot." In *Mankind at the Crossroads*, East wrote that "the negro race as a whole is possessed of undesirable transmissible qualities both physical and mental, which seem to justify not only a line but a wide gulf to be fixed permanently between it and the white race" (1928, 133). Blacks and whites were seen to represent too wide a racial cross; hybridization would only "break apart those compatible physical and mental qualities which have established a smoothly operating whole in each race by hundreds of generations of natural selection" (1928, 253). Disharmonies would surely result.

That was also the argument advanced by Charles Davenport in "The Effects of Race Intermingling." Davenport claimed that a well-established race is probably well-suited to its specific conditions, with its parts and functions "harmoniously adjusted." Thus the Scotch are tall, with internal organs well adapted for their large frames. Southern Italians, in contrast, have viscera well adjusted to their short frames. The hybrids of a cross between them could yield children similar to the parental types but also those with internal organs inadequate for their large frames and with organs too large for their small frames. These incompatibilities may be mental and temperamental as well. Thus one often finds in mulattoes ambition and drive combined with low intelligence, so that the hybrid is unhappy, dissatisfied with his fate, and rebellious. "To sum up, then," wrote Davenport, "miscegenation commonly spells disharmony. . . . A hybridized people are a badly put together people and a dissatisfied, restless, ineffective people." America would thus seem to be in trouble, for it was experiencing hybridization "on the greatest scale that the world has ever seen" (1917a, 365–67).

These ideas were expanded in his famous 1928 study (conducted with Morris Steggerda) *Race Crossing in Jamaica.* Davenport reported that crosses among a number of races, such as Canadians and Indians, or Chinese and Hawaiians, worked quite well. He even toyed with the idea that black-white mixing was desirable since it provided one way of injecting new variation into the system. Moreover, the Davenport-Steggerda study seemed to show that blacks had many virtues *complementary* to those of whites. Blacks, he thought, were more appreciative of music, were resistant to some diseases, suffered little tooth decay, and were better at arithmetic. If these traits were heritable (as was generally assumed), there were apparent benefits to crossing. Davenport considered whether it might not be desirable to mix blacks and Englishmen (who were sadly deficient in musical capacity).

In the end, he rejected the idea, having concluded that hybridization would produce a disproportionately large number of ineffective persons, the result of disharmonies produced by very wide crosses. In theory, it would be possible to combine crossing with a rigorous program of selection, thus culling the lower half of the hybrid population. But this scheme is impractical; we do not have the necessary control of matings. Furthermore, he argued, race-crossing would be a bad idea in any case, for "there exists in mankind a strong instinct for homogeneity." Davenport admonished his readers that

> a homogeneous group of white people will always be led by its instincts to segregate itself from Negroes, Chinese, and other groups that are morphologically dissimilar from themselves. We should consider the psychological, instinctive basis of this feeling. It is not sufficient merely to denounce it. It probably has a deep biological meaning and so long as it exists, so long we should be led to follow it as a guide if we are to seek to establish a commonwealth characterized by peace and unity of ideals. (1928, 238)

William E. Castle, another distinguished geneticist, fiercely attacked the study (and the use made of it by H. S. Jennings) in 1930. Castle insisted there was no evidence that race-crossing produced disharmonies; indeed, that Davenport and Steggerda's own data did not support their interpretation. He closed his *Science* article with the following prophecy:

We like to think of the Negro as inferior. We like to think of Negro-white crosses as a degradation of the white race. We look for evidence in support of that idea and try to persuade ourselves that we have found it even when the resemblance is very slight. The honestly made records of Davenport and Steggerda tell a very different story about hybrid Jamaicans from that which Davenport and Jennings tell about them in broad sweeping statements. The former will never reach the ears of eugenics propagandists and Congressional committees; the latter will be with us as the bogey men of pure-race enthusiasts for the next hundred years. (Castle 1930, 605–6).

Not surprisingly, Castle's articles were often invoked by antiracialists, as were other studies, such as the 1928 reports by Leslie Dunn and A. M. Tozzer on race-crossing in Hawaii and by Melville Herskovits on black-white crossing in the United States, that also failed to find evidence of disharmonies. But Castle's dispute with Davenport involved a scientific point, not the course of social policy. Castle was as strongly opposed as Davenport to miscegenation, although on social rather than biological grounds. In his textbook, he argued that "human racial crossing in general is a risky experiment, for it interferes with social inheritance, which after all is the chief asset of civilization." Black-white mating in particular would disturb families, schools, churches, and other social institutions. Crosses between such "widely separated branches of the human family" were therefore best avoided (Castle 1927, 332–33).

In an essay entitled "The Biological Effects of Race Mixture," the University of California zoologist Samuel J. Holmes reviewed the scientific evidence as to whether race-mixing was good or bad. He found existing studies to be superficial and marred by prejudice. Notwithstanding the claims of many geneticists, the studies did not show that hybridization produced disharmonies. Holmes nevertheless concluded that the crossing of races "on different mental levels" should be avoided, since they lowered the quality of the superior race. Indeed, there should be a uniform law, adopted by all states, barring the marriage of whites with blacks, Indians, and Chinese (Holmes 1923, 227). Even the crossing of "equivalent" races should probably be avoided, although there was no evidence showing it to be harmful: "The inheritance of a superior race is a very precious possession to be conserved at all costs," Holmes argued. We should not be taking leaps in the dark. *"The argument from ignorance should not be used to defend race crossing because we cannot prove that it is bad; it should be used rather to counsel caution because we do not know that it is not bad"* (1923, 223).

In their textbook, Popenoe and Johnson likewise noted that "the literature on the biology of race crossing is remarkable for [its]

abundance of preconceived ideas and scarcity of any real evidence"
(1933, 287). That did not stop them from promoting both
antimiscegenation and restrictive immigration laws. Indeed, they felt
that the Johnson-Reed Act did not go far enough. The preference it
"assigned to Nordic or partly Nordic peoples has given rise to a highly
emotional and sometimes bitter controversy," they noted, "particularly
from some non-Nordic persons who fancied that the preference given
to Nordics was invidious" (1933, 292). But they thought the law a sensible
attempt to prevent a dangerous threat to national unity. Alas, it did
not apply to the Philippines or to the Western Hemisphere, resulting
in unwanted immigration from Mexico, Puerto Rico, and the West Indies.
They urged that these defects be remedied (1933, 293–97; see also
Popenoe and Johnson 1918, 296–97).

In short, scientific studies did not prompt anyone inclined in another
direction to countenance race-mixing. Even those whose work led them
to consider hybrids generally superior to their parent races found other
reasons to exclude those crosses they found offensive. As in the debates
over segregation and sterilization, scientific criticisms had little impact
on attitudes toward miscegenation since the technical arguments against
it could always be replaced with social ones. Studies that seemed to
demonstrate the harmless or even beneficent effects of race-crossing
provided arguments only for those already disposed to make an
antiracialist case. That case became effective in a changed political
climate. As news of Nazi atrocities spread in the late 1930s and 1940s,
few geneticists wished to stress race differences (Provine 1986). Indeed,
they worked to distance their science as far as possible from the crimes
of the Nazi state. By the 1950s, genetics was said to have fatally
undermined racist arguments, for it proved that race-mixing was
scientifically sound. This reversal was prompted not by new studies
but by a change in values that led to a reevaluation of old ones. As we
saw earlier, scientific facts do not speak for themselves; they must always
be interpreted in light of social assumptions and goals. Revulsion at
fascist uses of genetics had produced a new reading of old evidence.
The scientific case against racism now seemed overwhelming.

From Eugenics to Human Genetics

T. H. MORGAN was a key figure in the development of modern genetics. Although generally apolitical and distrustful of radical movements, he was sufficiently disturbed by the passage of the Immigration Restriction Act to add a new chapter to the 1925 edition of *Evolution and Genetics* (Allen 1978, 21–22, 230–34). Here he argued that almost nothing was known about the causes of mental differences among individuals, much less among nations or races. Morgan also deplored the frequent confusion of nature and nurture, suggesting that much of the behavior associated with feeblemindedness was probably due to "demoralizing social conditions" rather than to heredity (Morgan 1925, 201). Until we know how large a role environment plays in human behavior, he advised, "the student of human heredity will do well to recommend more enlightenment on the social causes of deficiencies" rather than to adopt the eugenicists' panaceas. Morgan was particularly critical of those who asserted the superiority or inferiority of different races:

> If within each human social group the geneticist finds it impossible to discover, with any reasonable certainty, the genetic basis of behavior, the problems must seem extraordinarily difficult when groups are contrasted with each other where the differences are obviously connected not only with material advantages and disadvantages... but with traditions, customs, religions, taboos, conventions, and prejudices. A little goodwill might seem more fitting in treating these complicated questions than the attitude adopted by some of the modern race-propagandists. (1925, 207)

Morgan's argument was similar to one advanced more than a decade earlier by the Danish geneticist Wilhelm Johannsen. On the basis of experimental work with beans, Johannsen in 1909 had introduced a distinction between an organism's "genotype" (its internal set of hereditary factors) and its external appearance, or "phenotype." He argued that hereditary factors or "genes" (a term he also coined) do not themselves determine the phenotype, since the way in which genes are expressed depends on their context. Given the interaction of genes and environment, we cannot infer an organism's genetic makeup simply by looking at its physical appearance or behavior. By the same token, we cannot predict an organism's phenotype simply from a knowledge of its genes. Indeed, the same genotype may be expressed very differently in different environments. "Various hereditary malformations in poppies can be avoided if the earth is changed for the young plants," Johannsen explained, since there is a sensitive period of development during which the surrounding conditions have a decisive influence on the phenotype of the individual plants. He was quick to articulate the implications for eugenics: "There is no reason to assume that the weak and sickly would represent the genetically inferior stock—they could be individuals of the same [hereditary] value as children from higher social classes who are better cared for" (quoted in Hansen, in press).

Johannsen's antihereditarian message had little impact at the time, in Scandinavia or elsewhere. As we saw, geneticists in the United States remained silent as biological race differences were invoked in support of immigration restriction. Even H. S. Jennings, the one exception to this rule, did not then challenge the hereditarian arguments in favor of restriction. In his congressional testimony and writing at the time, Jennings argued that the law would fail to reduce the burden of defectiveness since it did not touch Ireland, which constituted "the chief source of degenerates" (Jennings 1923). Like Morgan, Jennings first criticized hereditarian doctrine only after restriction had passed (Jennings 1924, 1925). Moreover, after a brief flurry of activity, he had little to say on the subject of genetic differences until 1929—when he defended Charles Davenport's controversial study *Race Crossing in Jamaica* (Barkan 1992, 204–5). Limited as their activities were, during the 1920s Jennings and Morgan represented the only American critics of the dominant view that individual, group, and race differences in "talent and character" were largely attributable to differences in genes. In Britain, no geneticist challenged the hereditarian consensus. In this respect at least, Galton's program was wildly successful. However disparate their views on the morality and efficacy of specific eugenics proposals, virtually all geneticists agreed that heredity was the most important cause of social success and failure.

During the early 1930s, that situation began to change. The world economic crisis and rise of Nazism moved many geneticists politically to the left. These geneticists' increasing consciousness of economic inequalities and the threat of fascism made them much more sensitive to the class and racial biases that characterized the organized eugenics movement. The critiques of Morgan and Johannsen were increasingly taken up by others.

From Mainline to Reform Eugenics

In the United States, the most passionate critic of establishment eugenics was H. J. Miller, who had worked with Morgan (himself the most prominent critic of the eugenicists) at Columbia University. As a student in the 1910s, Muller had been attracted to Marxism. At the Third International Congress of Eugenics held in 1932 in New York, he delivered an impassioned speech entitled "The Dominance of Economics over Eugenics." Muller was not opposed to eugenics. Indeed, he began his speech with the remark "That imbeciles should be sterilized is of course unquestionable," although he also cautioned that the benefits of sterilization should not be exaggerated. Like many of his left-wing colleagues, Muller wished to reform, not destroy, eugenics. In his speech to the Congress and other writings of the 1930s, Muller argued that economic inequality masks genetic differences. Only at the very extremes of feeblemindedness and genius, he maintained, can we know with certainty that a particular genotype is deficient or superior. In a capitalist society characterized by "such glaring inequalities of environment as ours," genetic merit and environmental good fortune are necessarily confounded. Muller dismissed the view that social classes or races varied in genetic endowment, asserting that any differences are "accounted for fully by the known effects of environment" (Muller 1934, 141–42). In respect to individual mental and temperamental differences, genetic and environmental effects cannot be distinguished; hence capitalist societies are unable to make use of superior genotypes. Under socialism, he wrote, it would finally be possible "to recognize the best human material for what it is, and garner it from the great neglected tundras of humanity." Moreover, only in a socialist society would it be possible to exercise truly social control of reproduction (Muller 1934, 144).

Muller was one of many left-wing scientists who looked to the Soviet Union for inspiration. Like their radical colleagues in other fields, these scientists condemned economic exploitation and inequality. As scientists, they also saw themselves as engaged in a struggle against the forces of obscurantism. They were attracted by the Soviet commitment to materialism, to the rapid development of science, and to the promotion of a scientific outlook among the people. Left-wing geneticists assumed that this attitude would extend to questions of human breeding and that eugenics would finally receive a fair test. All the pieces for a successful experiment seemed to be in place. With equality of opportunity, the genetic wheat could finally be separated from the chaff. Because the Soviets placed the interests of society before those of individuals, a program to control human reproduction seemed congruent with the spirit of the new state. Moreover, if its leaders could be convinced of the desirability of eugenics, they had the power to implement it on a large scale.

In 1934, Muller, acting on his convictions, moved to the Soviet Union, bringing with him plans for a socialist eugenics. He sent a copy of his manuscript to Joseph Stalin, accompanied by a letter effusively praising Bolshevism and excoriating the race and class bias of capitalist eugenics. Muller's manuscript described a vast program for the artificial insemination of women with the sperm of men superior in intelligence, talent, and social feeling. "It is easy to show," he wrote, "that in the course of a paltry century or two . . . it would be possible for the majority of the population to become of the innate quality of such men as Lenin, Newton, Leonardo, Pasteur, Beethoven, Omar Khayyam, Pushkin, Sun Yat Sen, Marx (I purposely mention men of different fields and races), or even to possess their varied faculties combined" (Muller 1935, 113).

Like their peers in other countries, many Soviet geneticists had been enthusiastic eugenicists. After 1917, their concern focused on the ostensibly dysgenic effects of the Russian Revolution and the civil war and emigration of the intelligentsia that followed. They published genealogies of aristocratic and professionally eminent families and issued dire warnings about the consequences of biological decline. This nostalgia for the old social order left the eugenicists vulnerable to attack during the "cultural revolution" of the 1920s. At the same time, the Lamarckian doctrine of inheritance of acquired characters became increasingly popular. As sentiment turned against them, the Russian eugenicists turned either to the genetics of other organisms or to a transfigured eugenics, in which genealogies of outstanding proletarians replaced those of aristocrats (Graham 1977). Almost no one was fooled. Thus eugenics was already under attack when Muller arrived on the scene. His manuscript, published in 1935 as *Out of the Night*, was received much more enthusiastically in the West than it was in the Soviet Union. Notwithstanding his effusive praise for the Soviet leader, Stalin himself was unresponsive. In fact, Muller soon had to flee. But many readers outside of the Soviet Union were admiring. *Out of the Night* received an especially warm reception in Britain, where it was distributed by the Left Book Club. The British Communist newspaper, the *Daily Worker*, hailed it as a model for scientists while the Marxist journal *Science and Society* invited Muller to become a foreign editor (Paul 1984).

In Britain, many distinguished scientists who had been radicalized in the early 1930s shared Muller's worldview. In 1930, the mathematical geneticist Lancelot Hogben began to attack eugenicists for their slighting of environmental factors and blatant class bias. He was particularly critical of pedigree studies that claimed to show that feeblemindedness, pauperism, and crime were inherited on the grounds that they ran in families (Mazumdar 1992, 157–58).

Like Muller, who was a close friend, Hogben was not opposed to eugenics per se; on the contrary, he wanted to save it from its conservative champions. Hogben complained that eugenic propaganda had "been dominated by an explicit social bias which in England can only serve to render the eugenic standpoint unpalatable" to the working class. "By recklessly antagonizing the leaders of thought among the working classes," he charged, "the protagonists have done their best to make eugenics a matter of party politics with results which can only delay the acceptance of a national minimum of parenthood" (quoted in Mazumdar 1992, 150).

In the effort to purge eugenics of class bias, Hogben was joined by a number of eminent biologists such as Julian Huxley and J. B. S. Haldane. Historians have labeled as "reform eugenicists" those Anglo-American biologists who aimed to rid eugenics of its social prejudice and scientifically naive assumptions. There is certainly a real distinction to be drawn between scientifically sophisticated and politically left reformers like Muller and Hogben and the activists in the eugenics societies, who generally opposed egalitarian policies, dismissed environmental factors, and equated social position with genetic worth. However, the distinction between reform and mainline eugenicists should not be overdrawn. Not every conservative eugenicist was also scientifically naive. R. A. Fisher, a pioneer in the field of mathematical population genetics, represents a striking case in point. Fisher was right-wing in his politics and a mainstay of the establishment Eugenics Education Society. In this perspective, he would seem to be a "mainliner." But his worst enemies would grant that he was smart. Moreover, many reformers expressed class or race prejudices of their own (for example, Jennings's defense of the claim that black-white crossing leads to disharmonious combinations). While all reformers aimed to equalize opportunities, some assumed that workers as a whole were less intelligent and able than their social superiors (Paul 1984). But if the categories are rough (and in a few cases perhaps inapplicable), they do reflect real differences among eugenicists as a whole. Though the reform eugenicists were not always free of race or class bias, a gulf of scientific sophistication and political commitment separates Muller, Hogben, and Jennings from Madison Grant, Lothrop Stoddard, and Harry Laughlin.

By the mid-1930s, the reform movement touched even the centers of mainstream eugenics. The shift was most marked in the United States, where the Eugenics Record Office lost much of its influence. Davenport retired as director in 1934, and Harry Laughlin, by then an embarrassment to the Carnegie Institution, was retired in 1939; the ERO closed shortly thereafter (Allen 1986). The movement's center of gravity shifted to the American Eugenics Society (AES). Under the leadership of Frederick Osborn (1889–1981), who assumed control in 1927, it began to chart a more moderate course (a shift eased by resignations and retirements of some of its most extremist members). In particular, Osborn began to distance the society from racism. By the late 1930s he was insisting that there was little scientific evidence of innate differences among groups and warning against the ascription of superiority to any social class or race as a whole. He also invoked Sweden as a model, arguing that its housing subsidies, free day nurseries, maternal care, and other social welfare programs had induced members of the professional classes to have more children (Mehler 1988, 124–25, 270–71, 291–95). But these tentative steps toward reform were hardly underway before the movement was threatened by revelations of Nazi atrocities, which produced a backlash against eugenics of any kind. By the late 1940s, eugenics—whether reform or mainline—had fallen out of fashion with the public. The eugenics societies tried to adapt. They appointed distinguished scientists to their boards. They denounced Nazis and other bigots. And they increasingly turned their efforts toward the apparently neutral fields of birth control and human genetics.

As we saw in chapter 5, in the 1910s and 1920s eugenicists divided on the subject of contraception. Over time, as the futility of preventing the spread of birth control among middle- and upper-class women became increasingly evident, many more were converted. The biologist E. W. MacBride, vice-chairman of the Eugenics Education Society, noted in 1922 that the society had been reluctant to support birth control owing to the fear that its widespread practice "would prejudice the production of sufficient babies by the competent and far-seeing section of the community." However, he explained, a majority of its council now felt "that the damage which might have been apprehended from this source is already done; and whilst they feel that the logical remedy is the sterilization of the unfit, they recognize that public opinion in this country is not yet ready for such a measure" (MacBride 1922,

247). In the same year, the American Roswell Johnson argued similarly that "the present laws attempting to suppress Birth Control utterly fail to hold up the birth rate among superiors. When we turn to the inferior, we find it one of the most important means by which their relative super-fecundity is kept up" (Johnson 1922, 16). By the 1940s, it was obvious that efforts to bar contraception were fruitless at best and counterproductive at worst. Furthermore, changing public opinion had left eugenicists with few other options. The AES, under Osborn's leadership, began to aggressively promote the birth control cause. In 1952, Osborn was also appointed the first director of the Population Council, an organization funded by John D. Rockefeller III to promote what was now often called "family planning." (The label was invented by a public relations expert.) The council's efforts to find birth control methods suitable for mass use in the Third World ultimately led to the development of intrauterine devices (IUDs) and oral contraceptives (Reed 1978, 287–88).

Founding Human Genetics

Birth control was one of two major themes for the AES in the postwar period. The other was human genetics, broadly defined to encompass the heredity of behavior as well as disease. The connection between eugenics and human genetics is strikingly illustrated by the early history of the American Society of Human Genetics, which was founded in 1948. Five of the first six presidents—Lee Dice, C. P. Oliver, C. Nash Herndon, Sheldon Reed, and Franz Kallmann—served simultaneously as a member of the board of directors of the eugenics society. The exception was H. J. Muller, who refused to join the society for tactical reasons rather than out of an objection to eugenics.

From the start, human genetics was intertwined with—and sometimes indistinguishable from—eugenics. Much early work on the heredity of clinical diseases was pursued by eugenicists who were at least as interested in mental and moral traits. Charles Davenport, for example, worked simultaneously on the inheritance of Huntington's chorea, epilepsy, a cheerful temperament (which he thought dominant over a gloomy one), and "nomadism." The 1931 edition of *Human Heredity* by Erwin Bauer, Eugen Fischer, and Fritz Lenz (revised to include a discussion of the U.S. Army mental tests and their implications for human racial differences) described hundreds of anomalies and normal traits, some of which are today considered hereditary and some not. Like all human genetics textbooks prior to the 1950s, it discussed diseases, socially aberrant behaviors, and a host of mental and temperamental characteristics.

Its catalogue of traits included glaucoma, night and color blindness, cleft palate, various cancers, Parkinson's disease, and susceptibilities to rickets, goiter, hypertension, diabetes, gallstones, pernicious anemia, and tuberculosis, as well as schizophrenia, manic-depressive insanity, homosexuality, idiocy, genius, power of imagination, and talents for painting and sculpture, technical invention, and mathematics and science. Many capacities were assumed to be sex-linked. That "women are selected by nature mainly for the breeding of children and for the allurement of men" explains why they lag in mathematics and science but are more altruistic, insightful, and empathetic than men (Bauer et al. 1931, 599). It was supposed that each of these traits was hereditary because they ran in families. In the 1930s, human genetics relied almost exclusively on one method: the construction of pedigree charts.

According to Fritz Lenz, "The mental differences between human beings are not only much more extensive than the bodily differences, but are also enormously more important" (Bauer et al. 1931, 565). It would have been hard to find a geneticist who disagreed. Although Galton had assumed that eminent men tended to be especially vigorous, most eugenicists recognized that physical fitness did not correlate with intellectual achievement. Were it necessary to choose, they were in no doubt which should take precedence. Even in Germany, where the physically handicapped were subjected to sterilization and later murder, eugenicists emphasized mentality and behavior. Thus most sterilizations carried out under the 1933 law were for feeblemindedness, schizophrenia, and alcoholism. Only about one-tenth were for physical disorders (Proctor 1988, 107–8). The emphasis on behavior carried over into the postwar period. Even after the AES began to support work in medical genetics, mentality remained its primary concern. Osborn thought that the eugenics movement should not emphasize physical health. What really mattered, he argued, was a change in reproductive behavior by the intellectual elite (Osborn 1954, 2). "Eugenics is particularly interested in the psychological traits of intelligence and personality, because these traits are of major importance to civilization," he explained. "If there is justification for a broad eugenics movement, it is chiefly because of the part played by heredity in providing the necessary potentials for the development of high qualities of intelligence and personality" (Osborn 1951, 82).

But how to effect these changes in reproductive behavior? In the 1940s, the eugenics movement was as limited in its means as it was in its methods. Compulsory sterilization was out of fashion. Even tax subsidies to worthy families were politically unfeasible. The AES, along with other traditional patrons of eugenics, turned its attention to genetic counseling. Some individual geneticists, such as Davenport, had always provided "marriage advice" to those who sought their help. In the 1930s, formal clinics had been established in Germany and Denmark. A few were also established in the United States and Britain in the 1940s. Genetic counseling could play only a limited role in the effort to encourage breeding from the best. But in the perspective of the AES, something was better than nothing. In 1954, the first issue of the society's new journal announced a series on "heredity counseling." During the next four years, an article on this theme appeared in almost every issue. In fact, between 1954 and 1958, the journal published more articles on counseling than any other topic.

One of the first clinics in the United States was the Dight Institute for Human Genetics at the University of Minnesota, founded in 1941. Its patron was Charles Fremont Dight, who left his estate to the University of Minnesota to promote eugenics. Dight's will stipulated that the university "maintain a place for consultation and advice on heredity and eugenics and for rating of people, first, as to the efficiency of their bodily structure; second, as to their mentality; third as to their fitness to marry and reproduce" (Reed 1974, 3). President of the Minnesota Eugenics Society (as well as, briefly, Socialist alderman from Minneapolis's 12th ward) and author of many eugenic tracts, such as "Human Thoroughbreds, Why Not?" (1922), Dight lobbied vigorously for a state sterilization law and, after one was passed in 1925, for its extension to the noninstitutionalized. Like Marie Stopes, he also wrote to Hitler. Only instead of sending love poems, he praised Hitler's "plan to stamp out mental inferiority among the German people" (Dight 1933).

Dight was not the institute's only eugenicist sponsor. Charles M. Goethe, bank president and advocate of public playgrounds, also left much of his estate to the Dight Institute for its eugenic work. Indeed, with the exception of the American Cancer Society and the U.S. Public Health Service, virtually all the sponsors of human genetics had eugenic motivations. These patrons included the Rockefeller, Carnegie, Wenner-Grenn, McGregor, and Rackham foundations, the Pioneer and Commonwealth Funds, and various wealthy eccentrics, including Dight, Goethe, and textile magnate Colonel Wycliffe C. Draper, as well as the American Eugenics Society.

The most important patron was the Rockefeller Foundation, which in 1933 launched a program aimed at the rationalization and control of human behavior. Support for work in human genetics was one means toward that end. In the view of Alan Gregg, director of the division that funded work in human genetics, the field's potential was vast, since virtually all human characteristics were attributable to heredity.

Differences that had once been ascribed to the environment—in resistance to disease, intelligence, and temperament—were really due to genetic differences. Gregg charged that the public had an exaggerated faith in the efficacy of the environment and that their misplaced confidence had deplorable consequences. Educators wasted their efforts on hopeless material. Doctors looked to environment when heredity was to blame. In his view, "We have more to gain from wise matings than from $900,000 high schools." Gregg's mission was to convince physicians, educators, and the public at large of the power of heredity. With this end in view, from the 1930s to the 1950s his division funded many human genetics projects, including clinics for genetic counseling, in the United States, Scandinavia, Britain, and Germany (Paul 1991).

Of course, patrons do not always control the direction of programs they fund. In genetics, as in other disciplines, scientists sometimes take the money and run. In this case, both scientists and their sponsors generally agreed that human genetics should serve to improve the race. Most pioneers in the field of human genetics were active eugenicists. And they were not ashamed to say so. *Eugenics* was not yet a term of opprobrium among scientists and would not be until the 1960s. For example, Lawrence Snyder, one of the pioneers in human genetics in the United States, claimed that a knowledge of genetics might provide physicians with "the necessary information for setting up eugenic and euthenic programs for the protection of society" (1934, 706); Gordon Allen, that "clinical eugenics, in the form of genetic counseling, is already establishing itself as a proper part of medicine" (1955, 91); Tage Kemp, that "eugenics is a purely medical subject with the sole task of preventing disease. This form of eugenics is termed *genetic hygiene*" (1951, 297); and James Neel, that "what we are really discussing is a new eugenics, where I define eugenics simply as a collection of policies designed to improve the genetic well-being of our species" (Neel and Schull 1954, 256).

These geneticists, like many others, equated medical genetics with good eugenics. Even in the 1950s, few geneticists recoiled from the term. One exception was Lionel Penrose, who following World War II had been appointed Professor of Eugenics and head of the Galton Laboratory at University College, London. Penrose used the position—in 1964 renamed at his insistence Professor of Human Genetics—to challenge eugenics' scientific and social assumptions. An expert in the genetics of mental deficiency, he stressed the complexity of its causes and the modest impact of eugenical measures in reducing its incidence. He also argued that an index of a society's own health is its willingness to provide adequate care for those unable to care for themselves

(Penrose 1949, 238; see also Kevles 1985, 148–55, 251–52). But Penrose was nearly unique among geneticists in his wholesale rejection of eugenics. Most textbooks included sympathetic discussions of the subject, condemning past abuses but also taking for granted that reproduction was an act with social consequences and was thus legitimately a matter of social concern. The eugenics of the past, their authors conceded, was distorted by racial and class prejudice and simplistic scientific assumptions. But they insisted that eugenics has a rational core, which should be preserved. Some genes are unreservedly bad. Those that produce Tay-Sachs disease, muscular dystrophy, Huntington's chorea, and other serious conditions bring only misery to their bearers and unnecessary expense to society. The struggle to eliminate disease genes must be sharply demarcated from past policies that targeted ethnic and religious minorities and the poor.

Ultimately, however, the effort to distinguish a potentially good (medical) eugenics from the bad eugenics of the past proved unsuccessful. Public aversion to anything labeled eugenics (at least when called by that name) ultimately swamped the reform movement. Although Osborn was successful in attracting scientists to the society's board, its general membership declined steeply. In a concession to public sentiment, the society's journal, the *Eugenics Quarterly*, was renamed *Social Biology* in 1968. By then, it had been generally recognized that a successful eugenics program had to be called something else. Commenting on the new title, Osborn remarked that

> the name was changed because it became evident that changes of a eugenic nature would be made for reasons other than eugenics, and that tying a eugenic label on them would more often hinder than help their adoption. Birth control and abortion are turning out to be great eugenic advances of our time. If they had been advanced for eugenic reasons it would have retarded or stopped their acceptance. (Osborn 1977, 7)

Genetic Counseling

Under whatever rubric, medical genetics was of interest to the AES, as to other early patrons and most practitioners, because they wished to change the distribution of births in the population. Throughout the 1940s, 1950s, and 1960s, geneticists stressed the social character of their work. Whether or not they called it "eugenics," most believed that the heredity of the population should be a matter of public concern. Lee R. Dice, the first director of the University of Michigan's

Heredity Clinic, summarized the contributions to a panel on genetic counseling as follows:

> The panel stressed, and I think very wisely, the consideration of the heredity of future generations. We must give due concern to the possibility of eliminating, or, perhaps, of perpetuating, undesirable or desirable genes. We must not only be concerned with the particular family concerned, but also with whether or not harmful heredity may be continued or spread in our population. (Dice 1952a, 346)

For this reason, they also leaned to a directive approach in their counseling. Given their assumption that the counselor is responsible for the population as well as to the individual, most thought they should guide as well as inform their clients. Thus Clarence P. Oliver, the first director of the Dight Institute, asserted that "a geneticist should prevail upon some persons to have at least their share of children as well as show a black picture to those with the potentiality of producing children with undesirable traits" (Oliver 1952, 31). While some clinicians expressed optimism that, advised of their hereditary defect, clients would "nearly always follow their doctor's advice" (Kemp 1953, 241), most thought they needed at least a gentle push. Franz Kallmann believed that "persons requesting genetic advice cannot always be presumed to be capable of making a realistic decision as to the choice of a mate, or the advisability of parenthood, without support in the form of directive guidance and encouragement" (Kallmann 1958, 49). Curt Stern even anticipated the day when control would replace persuasion. "Natural selection will be superseded by socially decreed selection," he declared in his influential textbook:

> In the course of time ... the control by man of his own biological evolution will become imperative, since the power which knowledge of human genetics will place in man's hands cannot but lead to action. Such evolutionary controls will be world wide in scope, since, by its nature, the evolution of man transcends the concept of unrestricted national sovereignty. (Stern 1949, 603)

Most geneticists believed that it was what C. Nash Herndon called the "total genetic potential"—and not the single gene—that mattered (Herndon 1954, 66). Thus, if a couple were smart and responsible, it might be desirable if they reproduced, even at the risk of transmitting a genetic disease. Herndon thought that if a couple's "intellectual endowments and general genetic background" were far above the norm, "their reproduction might be advantageous to society as a whole, offsetting the disadvantage of the possible continuation of the defective gene"

(Herndon 1952, 335). Dice likewise believed that genetic counselors should consider more than the risk of medical disease. In giving advice, he argued, the geneticist should take into account "mental ability, and social worth in addition to hereditary defects" (Dice 1952b, 6).

A few geneticists argued that counseling should serve only the client. For example, Sheldon Reed, who coined the term "genetic counseling" in 1947, advocated neutrality in respect to reproductive decisions. "We try to explain thoroughly what the genetic situation is," he wrote in *Counseling in Medical Genetics*, "but the decision must be a personal one between the husband and wife, and theirs alone" (Reed 1955, 14). Reed felt that counselors should respect the opinions of those whose lives were directly affected by the decision to bear a child. Like other geneticists who objected to directive counseling, however, he also assumed that those individuals concerned enough about their future children to consult a counselor were usually well above average in the attributes that matter most: their intelligence and sense of responsibility. The value of transmitting these traits outweighed the risk of most diseases (Reed 1954, 48–49). Similarly, James Neel and William Schull explained that they did not attempt to influence families' reproductive decisions, noting "that it is often the more responsible and able members of society who are most impressed by the occurrence of abnormality in a child of theirs and most receptive to the possibility of altering their reproductive behavior accordingly" (1954, 308). Harold Falls also thought that those who seek counseling show "a greater degree of intelligence and social and moral responsibility on their part than is true of a large proportion of our population. Such parents, perhaps, should actually be encouraged to have children (anticipating transmission of superior qualities) providing the gene to be transmitted does not impose too serious a handicap on the affected child" (1959, 99).

Some clinicians also assumed that their clients did not need direct guidance to make the right choice. "From my experience in giving advice about heredity to families in all walks of life I can affirm that every parent desires his children to be free from serious handicap," wrote Dice. "If there is known to be a high probability of transmitting a serious defect, it would be an abnormal person indeed who would not refrain from having children" (1952b, 2). Counseling would automatically serve the interests both of individuals and of society. Given adequate information, the kinds of (middle-class) people who availed themselves of genetic services would make rational decisions. Thus James Neel, who argued consistently for nondirective counseling, believed that "once the principle of parental choice of a normal child is established, it seems probable that in large measure the parental desire for normal children can be relied on to result in the purely voluntary elimination of affected fetuses" (1970, 820–21; see also 1994, 361). Or, as Reed remarked:

> If our observation is generally correct, that people of normal mentality, who thoroughly understand the genetics of their problems, will behave in the way that seems correct to society as a whole, then an important corollary follows. It could be stated as a principle that the mentally sound will voluntarily carry out a eugenics program which is acceptable to society if counseling in genetics is available to them. (1952, 43)

In reality, genetic counseling could have only a trivial impact on the population as a whole. From a eugenical standpoint, it was thus inconsequential. Most disease genes are both recessive and extremely rare (much rarer than "feeblemindedness" was once thought to be). Since the number of normal carriers vastly outruns those affected, hundreds or even thousands of years would be required to substantially reduce the incidence of any genetic disease—even if all affected individuals could be prevented from breeding. Of course, no one suggested such a policy in respect to genetic diseases. The postwar reaction to Nazi atrocities made direct methods to control reproduction politically unfeasible. In any case, drastic measures hardly seemed rational given the slowness both of the potential spread and of the selection against extremely rare genes. Moreover, while only a small number of people may be affected with a *particular* genetic disease, many possess a hereditary defect of some kind—and they are found in every ethnic group, race, and class. Programs of segregation and sterilization had been targeted at "others." But genetic disease affects whites equally with blacks and the middle class with the poor. Under these conditions, compulsory measures had little popular appeal. Geneticists could at most offer advice and hope that their clients complied.

In fact, few sought their counsel. Before the 1970s, clinics had little to offer. Those who turned to geneticists for advice (typically parents who already had a child affected with a genetic disorder or who were anxious about transmitting a trait that ran in the family) were confronted with a stark choice based on often vague estimates of risk. Until abortion was legalized in the United States by the 1973 Supreme Court decision in *Roe v. Wade* and in Britain by a 1967 Act of Parliament, the only legal way to avoid genetic risk was not to reproduce. But the right to terminate a pregnancy would have had little impact in the absence of methods for detecting genetic disorders during pregnancy. In the 1960s, the first such method—amniocentesis—was developed, and by the mid-1970s, it had become a routine part of clinical practice. The convergence of prenatal diagnosis and legalization of abortion produced explosive growth in the field of genetic counseling. This rise in the number of clients and counselors was accompanied by a dramatic shift in the ethos of the field.

One might say that geneticists like Reed and Neel were ahead of their time. In the 1950s, their arguments for nondirective counseling fell on (mostly) deaf ears. That would soon change. Other forces were building that would ultimately have a profound effect on the ways people thought about the relations between the client, the counselor, and the larger society. Primary among these was the remarkable transformation in public attitudes toward reproductive responsibility that took place in the 1960s and 1970s. Until then, it was taken for granted that society had a legitimate interest in who reproduced. By the mid-1970s, it was equally taken for granted that society had no interest in the matter. Within two decades, reproduction was transformed from a public to a private concern.

From Reproductive Responsibility to Reproductive Autonomy

In the 1927 case of *Buck v. Bell*, upholding the Virginia sterilization statute, Justice Holmes had claimed: "We have seen more than once that the public welfare may call upon the best citizens for their lives. It would be strange if it could not call upon those who already sap the strength of the State for these lesser sacrifices, often not felt to such by those concerned, in order to prevent our being swamped with incompetence. . . . The principle that sustains compulsory vaccination is broad enough to cover cutting the fallopian tubes." In the 1972 case of *Eisenstadt v. Baird*, by contrast, the Supreme Court held that "if the right of privacy means anything, it is the right of the *individual*, married or single, to be free from unwarranted governmental intrusion in matters so fundamentally affecting a person as the decision whether to bear or beget a child." The shift in the law both reflected and reinforced other broad trends in the culture, including respect for patient rights in medicine and, especially, the resurgence of feminism. In the 1970s, a woman's right to control her body became a rallying cry of the feminist movement. Reproductive autonomy soon became a dominant value. The social movements that produced this sea-change in values are obviously of immense cultural importance and historical interest. Our concern here, however, is with their effect on the history of human genetics. When reproduction is viewed as a public matter, the role of the genetic counselor will be construed in one way; when viewed as a private matter, in quite another.

Within genetic counseling, concern for the future of the population was replaced by concern for the welfare of individual families, as defined by the families themselves. That change reflected events specific to the field as well as general trends in the culture. Most genetic counselors were originally male research-oriented Ph.D.s in genetics and, somewhat later, physicians. Counseling was usually provided as a sideline to other work. The first master's level program for professional counselors in the United States was established only in 1969. In a number of ways, the new counselors were quite different from their predecessors. All but a handful were women, who generally place a higher value on reproductive autonomy than do men (Wertz and Fletcher 1989, 221–41). The new counselors were also trained in "client-centered" therapy, which stresses the counselor's role in clarifying the client's own feelings. Thus factors specific to the field intersected with broad social forces to produce a strong noninterventionist stance.

By the late 1970s, there existed near-consensus on the aim of genetic counseling: "to inform, to educate, to convey value-free facts and probabilities about genetic conditions, perhaps even to deal with psychologic problems, but never to advise and counsel" (Twiss 1979, 201). Of course theory and practice may sometimes diverge. But at least in North America, the shift in ethos was dramatic. In 1968, the biochemist Linus Pauling could propose that all young people be tested for the presence of the sickle-cell and other deleterious genes and that a symbol be tatooed on the foreheads of those found to be carriers (Pauling 1968). It did not occasion much comment. The same proposal only a decade later would have provoked an uproar.

Genetic services are now everywhere justified as increasing the choices available to women. Counselors are said to serve their clients, not society. This perspective is codified in the 1991 Professional Code of Ethics of the National Society of Genetic Counselors, which defines the counselor-client relationship as "based on values of care and respect for the client's autonomy, individuality, welfare, and freedom" (Bartels et al. 1993, 170). But this shift in ethos may have come too late to matter, for genetic testing has expanded much more rapidly than the number of genetic counselors.

The past decade has witnessed a vast expansion in the genetic services offered to women. Less invasive forms of prenatal diagnosis—such as chorionic villus sampling, ultrasound, and (experimentally) the recovery of fetal cells from maternal blood—have become available.

So have predictive tests for an increasingly wide range of disorders in both fetuses and adults. In the 1970s, Down syndrome was the only major condition that could be detected in utero. Today, dozens of fetal anomalies, some rare and some quite common, are detectable. Tests also now exist to predict which individuals will bear offspring with genetic disorders. Persons with the dominant gene for Huntington's chorea, polycystic kidney disease, and forms of colon and breast cancer, among other late-onset disorders, can now be identified before they suffer symptoms and make reproductive decisions. Carriers of a number of recessive genes, such as those that produce Tay-Sachs disease, sickle-cell anemia, and cystic fibrosis, can also be identified.

The most rapid growth has been in the area of prenatal diagnosis. Today, many millions of women are faced with the choice of whether to have or refuse a genetic test. There are only about a thousand professional genetic counselors in the United States. Most women therefore learn of their choices from physicians, usually obstetricians or gynecologists (the situation that has always prevailed in Europe). Doctors are trained to be directive. Moreover, the fear of malpractice suits provides a strong incentive to promote the use of tests. If a woman refuses a test, gives birth to a child with a birth defect, and sues, how does a doctor prove that the refusal was well informed? It is legally safer if she takes the test. Thus the Department of Professional Liability of the American College of Obstetricians and Gynecologists (ACOG) alerted the organization's members to a need to inform *all* pregnant patients of the availability of screening for neural tube defects. The physicians were told that it was imperative that they inform every pregnant patient "of the advisability of this test and that your discussion about the test and the patient's decision with respect to the test be documented in the patient's chart." The rationale for the alert was legal, not medical: to provide the doctor with the most effective defense in the event of a malpractice suit. Indeed, ACOG had concluded that routine screening was of dubious value to the patient and should not be implemented in the absence of appropriate counseling and follow-up services (Annas and Elias 1985). Thus legal considerations spur doctors to promote the use of tests—and the doctor's attitude is the most important factor in determining whether or not women agree to be tested (Bekker et al. 1993).

Doctors are not the only actors with an interest in the expansion of genetic services. Biotechnology companies want to make money by selling predictive tests. Governments and insurers want to save money by having people use them. Policy makers typically think in terms of economic cost and benefit. The plus side of the cost-benefit ledger is represented by the number of terminations achieved for specific fetal conditions. The more women screened and affected fetuses aborted, the more efficient is the genetic service. Hence a cost-benefit rationale for these programs provides an incentive to expand genetic testing and to maximize abortions of fetuses affected with expensive disorders (Clarke 1990).

In the 1970s, the U.S. government played a major role in promoting amniocentesis. As the assistant secretary of the Department of Health, Education, and Welfare said at the time: "One of the main points of emphasis in the Department of HEW is prevention of disability. By focusing on prevention we increase the resources available for other programs. Few advances compare with amniocentesis in their capability for prevention of disability" (Cooper 1975, 2). Given the bitterness of the current debate over abortion, he would surely not be so open today. But the state still provides genetic services with the aim of saving money. As the philosopher Arthur Caplan notes, "When the state of California offers [a test] to all pregnant women it does so in the hope that some of those who are found to have children with neural tube defects will choose not to bring them to term; thereby, preventing the state from having to bear the burden of their care" (Caplan 1993, 159).

Conclusion

Policy makers generally assume that individual and social interests are congruent, that families will act "rationally." (Thus policy analyses of screening programs usually assume that all fetuses identified as abnormal will be aborted [Faden et al. 1987, 3].) As we have seen, the assumption that normal people will do what they can to avoid the birth of children with disabilities has a long history. It informed the views of Margaret Sanger, Havelock Ellis, Frederick Osborn, and Sheldon Reed. They argued that, when it came to the normal population, there was no need for compulsion. The race would be improved by the voluntary actions of poor women who wanted to limit their births. By the 1960s, it seemed politic to drop the old label for this program. As Frederick Osborn wrote in 1968, "Eugenic goals are most likely to be attained under a name other than eugenics" (1968, 25). Of course eugenics and reproductive choice are congruent only if families ordinarily make the "right" decisions.

Subtle pressures to make the "right" choice are what many people have in mind when they characterize contemporary genetic medicine as a form of eugenics. Of course many women welcome the opportunities to learn more about their fetus and to act on the results. But some women also feel that they have no realistic alternatives to the decision to be tested or to abort a genetically imperfect fetus. Of course they are under no legal coercion. But they may nevertheless feel pressured to be tested and avoid having children with disabilities—by their doctors, who fear being sued if the child is born with a genetic disorder, by anxiety over the potential loss of health or life insurance, by their inability to bear the enormous financial costs of caring for a severely disabled child, or by the lack of social services (even with national health insurance) for handicapped children.

Even if these pressures were eliminated, concerns about eugenics would not disappear. Thus some have argued that eugenics is being revived by our increased ability to choose the kind of children we want. "When eugenics reincarnates this time," warns the sociologist Troy Duster, "it will not come through the front door, as with Hitler's *Lebensborn* project. Instead, it will come by the back door of screens, treatments, and therapies" (Duster 1990, x; see also Wright 1990). Duster and others worry less about state coercion than citizen *demands*. Abby Lippman, a Canadian critic of genetic technologies, thinks that eugenics is already back. In her view, all prenatal diagnosis is ipso facto eugenics because it necessarily involves the systematic selection of fetuses. She notes: "Though the word 'eugenics' is scrupulously avoided in most biomedical reports about prenatal diagnosis, except where it is strongly disclaimed as a motive for intervention, this is disingenuous. Prenatal diagnosis presupposes that certain fetal conditions are intrinsically not bearable" (Lippman 1991b, 24–25).

Of course most medical geneticists strongly disagree. They tend to insist on a narrow definition of eugenics; for them, eugenics implies state interference with reproductive decisions. Such definitions sharply distinguish eugenics from medical genetics. The following distinction is typical: "Eugenics presumes the existence of significant social control over genetic and reproductive freedoms. Genetics does not require any special control over genetics or reproductive freedom" (Ledley 1991). However, a definition that requires coercion leads to seemingly absurd conclusions: that Frederick Osborn was not a eugenicist, or Havelock Ellis, or H. J. Muller, or even Francis Galton.

But it is certainly understandable that geneticists repudiate any suggestion that their activities constitute a form of eugenics. We have come to identify eugenics with the most terrible parts of its history. When we think of eugenics, it is usually not Margaret Sanger or Havelock Ellis who comes to mind but Madison Grant or Adolf Hitler. We do not think of free love and birth control but of compulsory sterilization or euthanasia. Eugenics evokes the image not of Denmark but of

Germany. Indeed, over every contemporary discussion of eugenics falls the shadow of the Third Reich. No wonder geneticists resist the label. To call their enterprise "eugenics" is thereby to condemn it.

We began this book with a question: has eugenics returned in the compassionate guise of medical genetics? From a historical standpoint, the answer would seem to be yes. But historical analysis is unlikely to end debate on this question, since the direction of public policy is (or is thought to be) at stake. For both advocates and critics of genetic medicine, the argument is thus political as well as historical. Or to put it somewhat differently, for many participants, history is a weapon in a war over social policy.

But if history will not settle this debate, it may explain its bitterness. It also alerts us to trends that ought to concern advocates as well as critics of developments in genetic medicine. It reveals that, when motives are mixed, financial considerations tend over time to displace other values. One clear lesson from the history of eugenics is this: what may be unthinkable when times are flush may come to seem only good common sense when they are not. In the 1920s, most geneticists found the idea of compulsory sterilization repugnant. In the midst of the Depression, they no longer did. When asylums were first established in the mid-nineteenth century, the "feebleminded" were thought to be trainable, and their education was stressed. Later in the century they came to be viewed as objects of pity, in need of protection from an often cruel world. By the turn of the twentieth century, they were perceived as a social menace and drain on the public purse. Over time, noble sentiments came increasingly to clash with economic demands. Charitable impulses gave way to utilitarian practices, and the economic value of the inmates' work came more and more to be stressed (Wolfensberger 1975, 31). But despite their superintendents' best efforts, the asylums never achieved self-sufficiency. During the world economic crisis of the 1930s, they everywhere came to be viewed as an unnecessary burden on society. And segregation gave way to compulsory sterilization.

Writing of American prison reformers in the Progressive Era, the historian David Rothman notes that they were motivated by both humanitarian and administrative concerns. But these interests were rarely congruent. "In the end," he wrote, "when conscience and convenience met, convenience won. When treatment and coercion met, coercion won" (1980, 10). As we have seen, the history of eugenics is also a story of diverse motivations. Advocates of birth control argued that contraception would increase reproductive choice for the normal population, thereby reducing the misery of workers and expense to the state. The same mix of motives is evident today. The expansion of genetic tests is defended variously on the grounds that tests increase the reproductive choices available to women and that they reduce the incidence of genetic disease, thus relieving both individual suffering and the burden on the public purse. The history of eugenics tells us which of these values is likely to triumph if and when they conflict.

History also illuminates a deep tension in the way we think about eugenics today. Popular discussions of eugenics focus on compulsory sterilization and Nazi breeding programs, that is, on brutal measures of state control. These accounts teach us to be wary of the prospect of genetics in the hands of the state. They therefore reinforce the commitment to reproductive autonomy. Indeed, that women have a right to control their bodies is often the explicit moral of these accounts. But many women who asserted that same right in the 1920s were themselves eugenicists. That may seem less paradoxical when we realize that eugenics represented a way of coping with human differences. The state was only one mechanism for dealing with the physically or mentally disabled. Many eugenicists thought that the job of ridding the world of the "unfit" could be as easily—or even better—carried out by individuals themselves. They only needed to be educated and given the tools for the job.

As a story of destructive state power, the history of eugenics teaches one lesson. As a story of attitudes toward people with disabilities, it teaches another. The first reinforces the view that reproduction should remain a private affair. The second leads us to reflect on attitudes toward the disabled, those whom Margaret Sanger as well as Adolf Hitler thought "should never have been born." In this second story, acts are not benign simply because their agents are private citizens. Indeed, if we insist on absolute reproductive autonomy we must accept the use of genetic technologies to prevent the birth of those who are unwanted for any reason: that they will be the "wrong" gender, or sexual orientation, or of short stature, or prone to obesity, or. . . . Used this way, medical genetics will surely reinforce a host of social prejudices. A history of eugenics that is sensitive to its complexities alerts us to the fact that genetic technologies present more than one kind of danger—and that if we are not very careful, we may avoid one only to court another.

Bibliography

Adams, Mark B. 1990a. "Eugenics in Russia." In *The Well-Born Science: Eugenics in Germany, France, Brazil, and Russia,* ed. M. Adams. New York: Oxford University Press.

Adams, Mark B. 1990b. "Toward a Comparative History of Eugenics." In *The Well-Born Science: Eugenics in Germany, France, Brazil, and Russia,* ed. M. Adams. New York: Oxford University Press.

Allen, Garland E. 1978. *Thomas Hunt Morgan: The Man and His Science.* Princeton: Princeton University Press.

Allen, Garland E. 1986. "The Eugenics Record Office at Cold Spring Harbor, 1910–1940: An Essay in Institutional History." *Osiris,* 2d ser., 2:225–64.

Allen, Garland E. 1989. "Eugenics and American Social History, 1880–1950." *Genome* 31:885–89.

Allen, Garland E. In press. "Eugenics." In *Encyclopedia of Genetics,* ed. J. Sapp. New York: Garland.

Allen, Gordon. 1955. "Perspectives in Population Genetics." *Eugenics Quarterly* 2:90–97.

Annas, George, and Sherman Elias. 1985. "Maternal Serum AFP: Educating Physicians and the Public." *American Journal of Public Health* 25 (December): 1374–75.

Bannister, Robert C. 1979. *Social Darwinism: Science and Myth in Anglo-American Social Thought.* Philadelphia: Temple University Press.

Barkan, Elazar. 1992. *The Retreat of Scientific Racism.* Cambridge: Cambridge University Press.

Barker, David. 1989. "The Biology of Stupidity: Genetics, Eugenics, and Mental Deficiency in the Inter-War Years." *British Journal for the History of Science* 22:347–75.

Bartels, D., B. LeRoy, and A. Caplan, eds. 1993. "Code of Ethics" (National Society of Genetic Counselors). In *Prescribing Our Future: Ethical Challenges in Genetic Counseling.* New York: Aldine de Gruyter.

Bauer, Erwin, Eugen Fischer, and Fritz Lenz. 1931. *Human Heredity.* 3d ed. Trans. E. and C. Paul. New York: Macmillan.

Bekker, Hilary et al. 1993. "Uptake of Cystic Fibrosis Testing in Primary Care: Supply Push or Demand Pull?" *British Medical Journal* 306 (June 12): 1584–86.

Bellamy, Edward. 1888. *Looking Backward.* New York: New American Library.

Bernard, William S. 1980. "Immigration: History of U.S. Policy." In *Harvard Encyclopedia of American Ethnic Groups,* ed. Stephen Thernstrom. Cambridge: Harvard University Press.

Blackmar, Frank W. [1897] 1988. "The Smoky Pilgrims." In *White Trash: The*

Eugenic Family Studies, 1877–1919, ed. N. H. Rafter. Boston: Northeastern University Press.

Boas, Franz. 1916. "Eugenics." *Scientific Monthly* 3:471–78.

Bock, Gisela. 1986. *Zwangsterilization im Nationalsozialismus*. Opladen: Westdeutscher.

Boris, Eileen. 1991. "Reconstructing the 'Family': Women, Progressive Reform, and the Problem of Social Control." In *Gender, Class, Race, and Reform in the Progressive Era*, ed. N. Frankel and N. S. Dye. Lexington: University of Kentucky Press.

Brigham, Carl C. [1923] 1975. *A Study of American Intelligence*. Reprint, with a foreword by R. M. Yerkes. Millwood, N.Y.: Kraus Reprint.

Broberg, Gunnar, and Mattias Tydén. In press. "Eugenics in Sweden: Efficient Care." In *Eugenics and the Welfare State: Sterilization Policy in Denmark, Sweden, Norway, and Finland*, ed. G. Broberg and N. Roll-Hansen. East Lansing: Michigan State University Press.

Bryan, William Jennings, and Mary Baird Bryan. 1925. *The Memoirs of William Jennings Bryan*. Philadelphia: John C. Winston.

Burleigh, Michael, and Wolfgang Wipperman. 1991. *The Racial State: Germany, 1933–1945*. Cambridge: Cambridge University Press.

Butler, Jon. 1990. *Awash in a Sea of Faith: Christianizing the American People*. Cambridge: Harvard University Press.

Caplan, Arthur L. 1993. "Neutrality Is Not Morality: The Ethics of Genetic Counseling." In *Prescribing Our Future: Ethical Challenges in Genetic Counseling*, ed. D. Bartels, B. LeRoy, and A. Caplan. Hawthorne, N.Y.: Aldine de Gruyter.

Carden, Maren Lockwood. 1969. *Oneida: Utopian Community to Modern Corporation*. Baltimore: Johns Hopkins Press.

Carson, John. 1993. "Army Alpha, Army Brass, and the Search for Army Intelligence." *Isis* 84:278–309.

Castle, William E. 1927. *Genetics and Eugenics*. 3d ed. Cambridge: Harvard University Press.

Castle, William E. 1930. Discussion. *Science* 71:603–6.

Castle, William E., J. M. Coulter, C. B. Davenport, E. M. East, and W. L. Porter. 1912. *Heredity and Eugenics*. Chicago: University of Chicago Press.

Chase, Allan. 1977. *The Legacy of Malthus: The Social Costs of the New Scientific Racism*. New York: Alfred A. Knopf.

Chesler, Ellen. 1992. *Woman of Valor: Margaret Sanger and the Birth Control Movement in America*. New York: Anchor Books.

Chesterton, G. K. 1922. *Eugenics and Other Evils*. London: Cassell.

Clarke, Angus. 1990. "Genetics, Ethics, and Audit." *Lancet* 335:1145–47.

Conklin, Edwin G. 1930. "The Purposive Improvement of the Human Race." In *Human Biology and Population Improvement*, ed. E. V. Cowdry. New York: Hoeber.

Coolidge, Calvin. 1921. "Whose Country Is This?" *Good Housekeeping* 72 (February): 13–14, 106, 109.

Cooper, Theodore. 1975. "Implications of the Amniocentesis Registry Findings." Unpublished Report. October.

Coren, Michael. 1993. *The Invisible Man: The Life and Liberties of H. G. Wells*. New York: Atheneum.

Cowan, Ruth Schwartz. 1977. "Nature and Nurture: The Inter-Play of Biology and Politics in the Work of Francis Galton." In *Studies in the History of*

Biology, ed. W. Coleman and C. Limoges. Baltimore: Johns Hopkins University Press.

Cravens, Hamilton. 1978. *The Triumph of Evolution: American Scientists and the Heredity-Evolution Controversy, 1900–1941*. Philadelphia: University of Pennsylvania Press.

Crook, Paul. 1994. *Darwinism, War and History*. Cambridge: Cambridge University Press.

Daniels, Roger. 1990. *Coming to America: A History of Immigration and Ethnicity in American Life*. New York: HarperCollins.

Darrow, Clarence. 1925. "The Edwardses and the Jukeses." *American Mercury* 6:147–57.

Darrow, Clarence. 1926. "The Eugenics Cult." *American Mercury* 8 (June): 129–37.

Darwin, Charles. [1859] 1964. *On the Origin of Species*. 1st ed. Reprint, with an introduction by Ernst Mayr. Cambridge: Harvard University Press.

Darwin, Charles. [1871] 1981. *The Descent of Man, and Selection in Relation to Sex*. Reprint. Princeton: Princeton University Press, 1981.

Darwin, Francis, and A. C. Secord, eds. 1903. *More Letters of Charles Darwin*. London: J. Murray.

Darwin, Leonard. 1928. *What Is Eugenics?* London: Watts.

Davenport, Charles B. 1909. "Report of Committee on Eugenics." American Breeders' Association.

Davenport, Charles B. 1911. *Heredity in Relation to Eugenics*. New York: Henry Holt.

Davenport, Charles B. 1912. "The Inheritance of Physical and Mental Traits of Man and Their Application to Eugenics." In *Heredity and Eugenics*, ed. W. E. Castle et al. Chicago: University of Chicago Press.

Davenport, Charles B. 1913. "Heredity, Culpability, Praiseworthiness, Punishment, and Reward." *Popular Science Monthly* 72 (July): 33–39.

Davenport, Charles B. 1915a. *The Feebly Inhibited: Nomadism, or the Wandering Impulse, with Special Reference to Heredity: Inheritance of Temperament*. Washington, D.C.: Carnegie Institution.

Davenport, Charles B. 1915b. Review. *Science* 42 (December 10): 837–38.

Davenport, Charles B. 1915c. "Review of H. H. Goddard, *Feeble-mindedness: Its Causes and Consequences*." *Science* 42: 837–38.

Davenport, Charles B. 1917a. "The Effects of Race Intermingling." *Proceedings of the American Philosophical Society* 46:364–68.

Davenport, Charles B. 1917b. "On Utilizing the Facts of Juvenile Promise and Family History in Awarding Naval Commissions to Untried Men." *Proceedings of the National Academy of Sciences* 3 (June): 404–9.

Davenport, Charles B. 1928. "Race Crossing in Jamaica." *Scientific Monthly* 27 (September): 225–38.

Davenport, Charles B., and Harry H. Laughlin. 1915. *How to Make a Eugenical Family Study*. Bulletin no. 13. Cold Spring Harbor, N.Y.: ERO.

Davenport, Charles B., H. H. Laughlin, David F. Weeks, E. R. Johnstone, and H. H. Goddard. 1911. *The Study of Human Heredity*. Bulletin no. 2. Cold Spring Harbor, N.Y.: ERO.

David, Paul R. 1933. "The 'Sterilization Spectre.'" *Journal of Heredity* 44:120–21.

Davies, Stanley P. 1930. *Social Control of the Mentally Deficient*. New York: Thomas Y. Crowell.

Deichmann, Ute. 1922. *Biologen unter Hitler: Vertreibung, Karrieren Forschung.* Frankfurt: Campus.

Deichmann, Ute, and Benno Müller-Hill. 1994. "Biological Research at Universities and Kaiser Wilhelm Institutes in Nazi Germany." In *Science, Technology, and National Socialism,* ed. M. Renneberg and M. Walker. Cambridge: Cambridge University Press.

Desmond, Adrian. 1989. *The Politics of Evolution: Morphology, Medicine, and Reform in Radical London.* Chicago: University of Chicago Press.

Dice, Lee R. 1952a. "Concluding Remarks" in "A Panel Discussion: Genetic Counseling." *American Journal of Human Genetics* 4 (December): 332–46.

Dice, Lee R. 1952b. "Heredity Clinics: Their Value for Public Service and Research." *American Journal of Human Genetics* 4 (March): 1–13.

Dice, Lee R. 1958. "The Structure of Heredity Counseling Services." *Eugenics Quarterly* 5:38–40.

Dight, Charles F. Letter to Adolph Hitler, August 1, 1933. Scrapbooks, Charles F. Dight Papers, Minnesota Historical Society.

Divine, Robert A. 1957. *American Immigration Policy, 1924–1952.* New Haven: Yale University Press.

Dugdale, Richard L. 1877. *"The Jukes": A Study in Crime, Pauperism, Disease, and Heredity.* New York: G. P. Putnam's Sons.

Durant, John. 1979. "Scientific Naturalism and Social Reform in the Thought of Alfred Russel Wallace." *British Journal for the History of Science* 12:31–57.

Duster, Troy. 1990. *Backdoor to Eugenics.* London: Routledge.

Dyer, Thomas G. 1980. *Theodore Roosevelt and the Idea of Race.* Baton Rouge: Louisiana State University Press.

East, Edward M. 1917. "Hidden Feeblemindedness." *Journal of Heredity* 8:215–17.

East, Edward M. 1927. *Heredity and Human Affairs.* Boston: Scribner's.

East, Edward M. 1928. *Mankind at the Crossroads.* Boston: Scribner's.

East, Edward M., and Donald F. Jones. 1919. *Inbreeding and Outbreeding: Their Genetic and Sociological Significance.* Philadelphia: J. B. Lippincott.

Ellis, Havelock. [1914] 1978. *The Task of Social Hygiene.* Farmingdale, N.Y.: Dabor Social Science Publications.

Faden, Ruth et al. 1987. "Prenatal Screening and Pregnant Women's Attitudes toward the Abortion of Defective Fetuses." *American Journal of Public Health* 77:1–3.

Falls, Harold. 1959. "Consideration of the Whole Person." In *Heredity Counseling,* ed. H. Hammons. New York: Hoeber-Harper.

Fernald, Walter E. 1912. "The Burden of Feeblemindedness." *Medical Communications of the Massachusetts Medical Society* 33.

Fichman, Martin. 1981. *Alfred Russel Wallace.* Boston: Twyane.

Fitzgerald, F. Scott. [1925] 1992. *The Great Gatsby.* Reprint, with preface and notes by M. J. Bruccoli. New York: Macmillan.

Forrest, D. W. 1974. *Francis Galton: The Life and Work of a Victorian Genius.* London: Elek.

Freeden, Michael. 1979. "Eugenics and Progressive Thought: A Study in Ideological Affinity." *Historical Journal* 22:645–71.

Frykman, Jonas. 1981. "Pure and Rational: The Hygienic Vision: A Study of Cultural Transformation in the 1930's 'The New Man.'" *Ethnologia Scandinavica* 26–63.

Gallup, George, Jr. and Jim Castelli. 1989. *The People's Religion: American Faith in the '90s*. New York: Macmillan.

Galton, Francis. 1865. "Hereditary Talent and Character." *Macmillan's Magazine* 12:157–66, 318–27.

Galton, Francis. 1869. *Hereditary Genius*. London: Macmillan.

Galton, Francis. [1874] 1970. *English Men of Science: Their Nature and Nurture*. Reprint, with an introduction by Ruth Schwartz Cowan. London: Cass.

Galton, Francis. 1883. *Inquiries into Human Faculty and its Development*. London: J. M. Dent and Sons.

Galton, Francis. 1908. *Memories of My Life*. London: Methuen.

Galton, Francis. [1909a] 1985. "Eugenics: Its Definition, Scope, and Aims." In *Essays in Eugenics*. Reprint. New York: Garland.

Galton, Francis. [1909b] 1985. "The Possible Improvement of the Human Breed, under the Existing Conditions of Law and Sentiment." In *Essays in Eugenics*. Reprint. New York: Garland.

Galton, Francis. [1909c] 1985. "Probability: The Foundation of Eugenics." In *Essays in Eugenics*. Reprint. New York: Garland.

Gelb, Steven A. 1985. "Myths, Morons, Psychologists: The Kallikak Family Revisited." *Review of Education* 11:255–59.

Gelb, Steven A. 1987. "Social Deviance and the 'Discovery' of the Moron." *Disability, Handicap, and Society* 2:247–58.

Gelb, Steven A. 1990. "Degeneracy Theory, Eugenics, and Family Studies." *Journal of the History of the Behavioral Sciences* 26 (July): 242–45.

Goddard, Henry H. 1911–12. "Ungraded Classes." *Report on Educational Aspects of the Public School Systems of the City of New York, Part II, Subdivision I, Section E*. New York: City of New York.

Goddard, Henry H. 1912. *The Kallikak Family: A Study in the Heredity of Feeble-Mindedness*. New York: Macmillan.

Goddard, Henry H. 1914. *Feeble-Mindedness: Its Causes and Consequences*. New York: Macmillan.

Goddard, Henry H. 1917. "Mental Tests and the Immigrant." *Journal of Delinquency* 2:243–77.

Goddard, Henry H. [1920] 1984. *Human Efficiency and Levels of Intelligence*. Reprint. New York: Garland.

Gökyiğit, Emel Aileen. 1994. "The Reception of Francis Galton's *Hereditary Genius* in the Victorian Periodical Press." *Journal of the History of Ideas* 27:215–40.

Gordon, Linda. 1990. *Woman's Body, Woman's Right: A Social History of Birth Control in America*. 2d ed. New York: Penguin.

Gould, Stephen Jay. 1981. *The Mismeasure of Man*. New York: W. W. Norton.

Gould, Stephen Jay. 1988. "On Replacing the Idea of Progress with an Operational Notion of Directionality." In *Evolutionary Progress*, ed. M. H. Nitecki. Chicago: University of Chicago Press.

Graham, Loren. 1977. "Science and Values: The Eugenics Movement in Germany and Russia in the 1920s." *American Historical Review* 82:1133–64.

Grant, Madison. 1916. *The Passing of the Great Race*. New York: Charles Scribner's Sons.

Greene, John. 1981. "Darwin as a Social Evolutionist." In *Science, Ideology, and World View: Essays in the History of Evolutionary Ideas*. Berkeley and Los Angeles: University of California Press.

Greg, William. 1868. "On the Failure of 'Natural Selection' in the Case of Man." *Fraser's Magazine* 68 (September): 353–62.

Gruber, Howard E., and Paul H. Barrett. 1974. *Darwin on Man: A Psychological Study of Scientific Creativity Together with Darwin's Early and Unpublished Notebooks.* Transcribed and ed. by Paul H. Barrett. New York: E. P. Dutton.

Haldane, J. B. S. 1939. "Heredity: Some Fallacies." *Science and Everyday Life.* London: Lawrence and Wishart.

Hall, G. Stanley. 1904. *Adolescence: Its Psychology and Its Relations to Physiology, Anthropology, Sociology, Sex, Crime, Religion, and Education.* 2 vols. New York: D. Appleton.

Haller, Mark H. 1985. *Eugenics: Hereditarian Attitudes in American Thought.* New Brunswick: Rutgers University Press.

Hansen, Bent Sigurd. In press. "Something Rotten in the State of Denmark: Eugenics and the Ascent of the Welfare State." In *Eugenics and the Welfare State: Sterilization Policy in Denmark, Sweden, Norway, and Finland,* ed. G. Broberg and N. Roll-Hansen. East Lansing: Michigan State University Press.

Hartl, Daniel L., and Vitezlav Orel. 1992. "What Did Gregor Mendel Think He Discovered?" *Genetics* 131 (June): 245–53.

Harwood, John. 1989. "Genetics, Eugenics, and Evolution." *British Journal for the History of Science* 22 (September): 257–65.

Herndon, C. Nash. 1952. "Statement" in "A Panel Discussion: Genetic Counseling." *American Journal of Human Genetics* 4 (December): 332–46.

Herndon, C. Nash. 1954. "Heredity Counseling." *Eugenics Quarterly* 1:64–66.

Hitler, Adolf. [1925] 1971. *Mein Kampf.* Trans. Ralph Manheim. Boston: Houghton Mifflin.

Hobhouse, Leonard. 1911. *Social Evolution and Political Theory.* New York: Columbia University Press.

Hogben, Lancelot. 1931. *Genetic Principles in Medicine and Social Science.* London: Williams and Norgate.

Holmes, Samuel J. 1923. *Studies in Evolution and Eugenics.* New York: Harcourt Brace.

Hunt, H. R. 1933. "Interest Increasing." *Journal of Heredity* 44:150–51.

Hunter, George William. 1914. *A Civic Biology: Presented in Problems.* New York: American Book.

Hutchinson, E. P. 1981. *Legislative History of American Immigration Policy, 1798–1965.* Philadelphia: University of Pennsylvania Press.

Jennings, H. S. 1923. "Undesirable Aliens." *Survey* 51 (December 15): 309–12, 364.

Jennings, H. S. 1924. "Heredity and Environment." *Scientific Monthly* 19 (September): 234–38.

Jennings, H. S. 1925. *Prometheus; or, Biology and the Advancement of Man.* New York: E. P. Dutton.

Jennings, H. S. 1927. "Health Progress and Race Progress: Are They Incompatible?" *Journal of Heredity* 18:271–76.

Jennings, H. S. 1930. *The Biological Basis of Human Nature.* London: Faber and Faber.

Johnson, Roswell H. 1922. "Eugenic Aspects of Birth Control." *Birth Control Review* 6:16.

Jones, Greta. 1986. *Social Hygiene in Twentieth-Century Britain.* London: Croom Helm.

Kallmann, Franz. 1958. "Types of Advice Given by Heredity Counselors." *Eugenics Quarterly* 5:48–50.

Kemp, Tage. 1951. *Genetics and Disease.* Copenhagen: Ejnar Munksgaard.

Kemp, Tage. 1953. "Genetic Hygiene and Genetic Counseling." *Acta Genetica et Statistica Medica* 4:240–47.

Kempton, J. H. 1933. "Bricks without Straw." *Journal of Heredity* 24:463–66.

Kennedy, David M. 1970. *Birth Control in America.* New Haven: Yale University Press.

Kevles, Daniel J. 1968–69. "Testing the Army's Intelligence: Psychologists and the Military in World War I." *Journal of American History* 55:565–81.

Kevles, Daniel J. 1985. *In the Name of Eugenics: Genetics and the Uses of Human Heredity.* Berkeley and Los Angeles: University of California Press.

Keynes, Richard Darwin. 1988. *Charles Darwin's Beagle Diary.* New York: Cambridge University Press.

King, Miriam, and Steven Ruggles. 1990. "American Immigration, Fertility, and Race Suicide at the Turn of the Century." *Journal of Interdisciplinary History* 20:347–69.

Koch, Lene. 1993. "The Biological Aspect." In *Rescue—43. Xenophobia and Exile.* Copenhagen: University of Copenhagen.

Koch, Lene. 1994. "Correspondences: On Danish Human Genetics and German Racial Hygiene in the 1930s." Unpublished ms.

Kottler, Malcolm. 1974. "Alfred Russel Wallace, the Origin of Man, and Spiritualism." *Isis* 65:145–92.

Kramnick, Isaac, and Barry Sheerman. 1993. *Harold Laski: A Life on the Left.* New York: Penguin.

Kühl, Stefan. 1994. *The Nazi Connection: Eugenics, American Racism, and German National Socialism.* New York: Oxford University Press.

Lagemann, Ellen Condliffe. 1992. *The Politics of Knowledge: The Carnegie Corporation, Philanthropy, and Public Policy.* Chicago: University of Chicago Press.

Larson, Edward J. 1991. "Belated Progress: The Enactment of Eugenic Legislation in Georgia." *Journal of the History of Medicine and Allied Sciences* 46:44–64.

Larson, Edward J. 1995. *Sex, Race, and Science: Eugenics in the Deep South.* Baltimore: Johns Hopkins University Press.

Laski, H. J. 1910. "The Scope of Eugenics." *Westminster Review* 174:25–34.

Laughlin, H. H. 1914. "Report of the Committee to Study and to Report on the Best Practical Means of Cutting Off the Defective Germ-Plasm in the American Population. I. The Scope of the Committee's Work." Bulletin no. 10A. Cold Spring Harbor, N.Y.: ERO.

LeConte, Joseph. 1879. "The Genesis of Sex." *Popular Science Monthly* 16 (December): 167–79.

Ledley, F. D. 1991. "Differentiating Genetics and Eugenics on the Basis of Fairness." *Proceedings of the Eighth International Congress of Human Genetics.* Supplement, *American Journal of Human Genetics* 49 (October): 325.

Lippman, Abby. 1991a. "Mother Matters: A Fresh Look at Prenatal Genetic Testing." *Issues in Reproductive and Genetic Engineering* 5:141–154.

Lippman, Abby. 1991b. "Prenatal Genetic Testing and Screening: Constructing Needs and Reinforcing Inequities." *American Journal of Law and Medicine* 17:15–50.

Love, Rosaleen. 1979. "'Alice in Eugenics-Land': Feminism and Eugenics in the Scientific Careers of Alice Lee and Ethel Elderton." *Annals of Science* 36:145–58.

Ludmerer, Kenneth M. 1972. *Genetics and American Society: A Historical Appraisal.* Baltimore: Johns Hopkins University Press.

MacBride, E. W. 1922. "British Eugenists and Birth Control." *Birth Control Review* 6:247.

McCormick, Richard L. 1990. "Public Life in Industrial America, 1877–1917." In *The New American History*, ed. E. Foner. Philadelphia: Temple University Press.

McLaren, Angus. 1990. *Our Own Master Race: Eugenics in Canada, 1885–1945.* Toronto: McClelland & Stewart.

Macrakis, Kristie. 1993. *Surviving the Swastika: Scientific Research in Nazi Germany.* New York: Oxford University Press.

Marchant, James. 1916. *Alfred Russel Wallace: Letters and Reminiscences.* New York: Harper and Brothers.

Marks, Richard Lee. 1991. *Three Men of the Beagle.* New York: Alfred A. Knopf.

Martin, Victoria Clafin Woodhull. [1891] 1974. "The Rapid Multiplication of the Unfit." In *The Victoria Woodhull Reader*, ed. M. Stern. Weston, Mass: M & S Press.

Mateer, Florence. 1913. "Mental Heredity and Eugenics." *Psychological Bulletin* 10 (June): 224–29.

Mazumdar, Pauline M. H. 1992. *Eugenics, Human Genetics and Human Failings: The Eugenics Society, Its Sources and Its Critics in Britain.* London: Routledge.

Mehler, Barry Alan. 1988. "A History of the American Eugenics Society, 1921–1940." Ph.D. diss., University of Illinois at Urbana-Champaign.

Morgan, T. H. 1925. *Evolution and Genetics.* 2d ed. Princeton: Princeton University Press.

Muller, H. J. [1934] 1984. "Dominance of Economics over Eugenics." In *A Decade of Progress in Eugenics.* Third International Congress of Eugenics, 1932. Reprint. New York: Garland.

Muller, H. J. [1935] 1984. *Out of the Night.* Reprint. New York: Garland.

Müller-Hill, Benno. 1988. *Murderous Science.* Trans. George R. Fraser. Oxford: Oxford University Press.

Muncy, Robyn. 1991. *Creating a Female: Dominion in American Reform, 1890–1935.* New York: Oxford University Press.

Neel, James V. 1970. "Lessons from a Primitive People." *Science* 170:815–21.

Neel, James V. 1994. *Physician to the Gene Pool: Genetic Lessons and Other Stories.* New York: John Wiley and Sons.

Neel, James V., and William Schull. 1954. *Human Heredity.* Chicago: University of Chicago Press.

Neuhaus, Richard John. 1990. *Guaranteeing the Good Life: Medicine and the Return of Eugenics.* Grand Rapids, Mich.: W. B. Eerdmans.

Ochsner, A. J. 1899. "Surgical Treatment of Habitual Criminals." *Journal of the American Medical Association* 32 (April): 867–68.

Oliver, Clarence P. 1952. "Human Genetics Program at the University of Texas." *Eugenical News* 37:25–31.

Osborn, Frederick. 1951. *Preface to Eugenics.* 2d ed. New York: Harper and Brothers.

Osborn, Frederick. 1954. Editorial. *Eugenics Quarterly* 1:2.

Osborn, Frederick. 1968. *The Future of Human Heredity*. New York: Weybright and Talley.

Osborn, Frederick. 1977. Transcript, Oral History Interview (July 10, 1974), Columbia University, New York.

Painter, Nell Irvin. 1987. *Standing at Armageddon: The United States, 1877–1919*. New York: W. W. Norton.

Paul, Diane B. 1984. "Eugenics and the Left." *Journal of the History of Ideas* 45:567–90.

Paul, Diane B. 1991. "The Rockefeller Foundation and the Origins of Behavior Genetics." In *The Expansion of American Biology*, ed. K. Benson et al. New Brunswick: Rutgers University Press.

Paul, Eden, 1917. "Eugenics, Birth-Control, and Socialism." In *Population and Birth-Control: A Symposium*, ed. Eden Paul and Cedar Paul. New York: Critic and Guide.

Pauling, Linus. 1968. "Reflections on the New Biology." *UCLA Law Review* 15:267–72.

Pearson, Karl. [1900] 1985. "National Life from the Standpoint of Science." *Eugenics Laboratory Lecture Series*. 2d ed. Reprint. New York: Garland.

Pearson, Karl. [1907] 1985. "On the Scope and Importance to the State of the Science of National Eugenics." *Eugenics Laboratory Lecture Series*, 2d ed. Reprint. New York: Garland.

Pearson, Karl. 1930. *The Life, Letters and Labours of Francis Galton*. Vol. 3A. Cambridge: Cambridge University Press.

Pendleton, Hester. [1871] 1984. *The Parent's Guide; or, Human Development through Inherited Tendencies*. Reprint. New York: Garland.

Penrose, Lionel S. 1949. *The Biology of Mental Defect*. London: Sidwick and Jackson.

Phillips, J. C. 1916. "Harvard and Yale Birth Rates." *Journal of Heredity* 7:565–69.

Pick, Daniel. 1989. *Faces of Degeneration: A European Disorder, c. 1848–c. 1918*. Cambridge: Cambridge University Press.

Pickens, Donald K. 1968. *Eugenics and the Progressives*. Nashville: Vanderbilt University Press.

Pittinger, Mark. 1993. *American Socialists and Evolutionary Thought, 1870–1920*. Madison: University of Wisconsin Press.

Popenoe, Paul. 1915. "Feeblemindedness." *Journal of Heredity* 6:32–36.

Popenoe, Paul, and Roswell H. Johnson. 1918. *Applied Eugenics*. New York: Macmillan.

Popenoe, Paul, and Roswell H. Johnson. 1933. *Applied Eugenics*. 2d ed. New York: Macmillan.

Proctor, Robert N. 1988. *Racial Hygiene: Medicine under the Nazis*. Cambridge: Harvard University Press.

Provine, William R. 1986. "Geneticists and Race." *American Zoologist* 26:857–87.

Punnett, R. C. 1917. "Eliminating Feeblemindedness." *Journal of Heredity* 8:464–65.

Radford, John P. 1991. "Sterilization versus Segregation: Control of the 'Feebleminded,' 1900–1938." *Social Science and Medicine* 33:449–58

Rafter, Nicole Hahn. 1988. Introduction. In *White Trash: The Eugenic Family Studies, 1877–1919*, ed. N. H. Rafter. Boston: Northeastern University Press.

Reed, James. 1978. *The Birth Control Movement in American Society: From Private*

Vice to Public Virtue. Princeton: Princeton University Press.

Reed, Sheldon. 1952. "Heredity Counseling and Research." *Eugenical News* 37:41–46.

Reed, Sheldon. 1954. "Heredity Counseling." *Eugenics Quarterly* 1: 47–51.

Reed, Sheldon. 1955. *Counseling in Medical Genetics.* Philadelphia: W. B. Saunders.

Reed, Sheldon. 1974. "A Short History of Genetic Counseling." Dight Institute Bulletin no. 14, 1–10.

Reilly, Philip R. 1991. *The Surgical Solution: A History of Involuntary Sterilization in the United States.* Baltimore: Johns Hopkins University Press.

Rogers, A. C., and Maud A. Merrill. [1919] 1988. "Dwellers in the Vale of Siddem." In *White Trash: The Eugenic Family Studies, 1877–1919*, ed. N. H. Rafter. Boston: Northeastern University Press.

Roll-Hansen, Nils. 1989. "Geneticists and the Eugenics Movement in Scandinavia." *British Journal for the History of Science* 22:335–346.

Roll-Hansen, Nils. In press. "Norwegian Eugenics: Sterilization as Part of a Progressive Program for Social Reforms." In *Eugenics and the Welfare State: Sterilization Policy in Denmark, Sweden, Norway, and Finland*, ed. G. Broberg and N. Roll-Hansen. East Lansing: Michigan State University Press.

Roosevelt, Theodore. 1926. "Marriage and Divorce." In *Works of Theodore Roosevelt: State Papers as Governor and President*, vol. 15. New York: Charles Scribner's Sons.

Rose, June. 1992. *Marie Stopes and the Sexual Revolution.* London: Faber and Faber.

Rosenberg, Charles E. 1976. *No Other Gods: On Science and American Social Thought.* Baltimore: Johns Hopkins University Press.

Ross, Edward A. 1901. "The Causes of Race Superiority." *Annals of the American Academy of Political and Social Science* 18 (July 1): 67–89.

Ross, Edward A. 1914. *The Old World in the New.* New York: Century.

Rossiter, Margaret W. 1982. *Women Scientists in America: Struggles and Strategies to 1940.* Baltimore: Johns Hopkins University Press.

Rothman, David J. 1980. *Conscience and Convenience: The Asylum and Its Alternatives in Progressive America.* Boston: Little, Brown.

Royal Commission on the Care and Control of the Feeble-Minded. 1908. "The Problem of the Feeble-Minded. An Abstract of the *Report*." London: HMSO.

Russett, Cynthia Eagle. 1989. *Sexual Science: The Victorian Construction of Womanhood.* Cambridge: Harvard University Press.

Rydell, Robert W. 1993 *World of Fairs: The Century-of-Progress Expositions*, Chicago: University of Chicago Press.

Sanger, Margaret. [1922] 1950. *The Pivot of Civilization.* Reprint. Elmsford, N.Y.: Maxwell Reprint.

Schneider, William. 1990. *Quality and Quantity: The Quest for Biological Regeneration in Twentieth-Century France.* Cambridge: Cambridge University Press.

Schuster, Edgar, and Ethel M. Elderton. [1907] 1985. *The Inheritance of Ability, Being a Statistical Study of the Oxford Class Lists and of the School Lists of Harrow and Charterhouse.* Reprint. New York: Garland.

Searle, G. R. 1976. *Eugenics and Politics in Britain, 1900–1914.* Leyden: Noordhoff International.

Selden, Steven. 1989. "The Use of Biology to Legitimate Inequality: The Eugenics Movement within the High School Biology Textbook, 1914–1949" In *Equity in Education*, ed. Walter Secada. New York: Falmer Press.

Sharp, Harry C. 1902. "The Severing of the Vasa Deferentia and its Relation to the Neuropsychopathic Constitution." *New York Medical Journal* 75 (March): 411–14.

Shaw, George Bernard. 1905. "Comment." In *Sociological Papers*, vol 1. London: Macmillan.

Shaw, George Bernard. 1928. *The Intelligent Woman's Guide to Socialism and Capitalism.* New York: Brentano's.

Shenton, James P. 1990. "Ethnicity and Immigration." In *The New American History,* ed. E. Foner. Philadelphia: Temple University Press.

Smith, J. David. 1985. *Minds Made Feeble: The Myth and Legacy of the Kallikaks.* Rockville, Md.: Aspen Systems.

Snyder, Lawrence. 1934. *The Principles of Heredity.* Boston: Heath.

Solomon, Barbara M. 1956. *Ancestors and Immigrants.* Cambridge: Harvard University Press.

Soloway, Richard A. 1990. *Demography and Degeneration: Eugenics and the Declining Birthrate in Twentieth-Century Britain.* Chapel Hill: University of North Carolina Press.

Stepan, Nancy L. 1991. *The Hour of Eugenics: Race, Gender, and Nation in Latin America.* Ithaca: Cornell University Press.

Stern, Curt. 1949. *Principles of Human Genetics.* San Francisco: W. H. Freeman.

Stoddard, Lothrop. [1923] 1984. *The Revolt against Civilization: The Menace of the Under Man.* Reprint. New York: Garland.

Stoler, Ann Laura. 1991. "Carnal Knowledge and Imperial Power: Gender, Race, and Morality in Colonial Asia." In *Gender at the Crossroads of Knowledge,* ed. M. di Leonardo. Berkeley and Los Angeles: University of California Press.

Stopes, Marie C. 1921a. *Radiant Motherhood: A Book for Those Who Are Creating the Future.* London: G. P. Putnam's Sons.

Stopes, Marie C. 1921b. *Verbatim Report of the Town Hall Meeting,* October 27. New York: Voluntary Parenthood League.

Tobey, Ronald C. 1971. *The American Ideology of National Science, 1919–1930.* Pittsburgh: University of Pittsburgh Press.

Tönnies, Ferdinand. 1906. Comment. In *Sociological Papers,* vol. 2. London: Macmillan.

Trent, James W., Jr. 1994. *Inventing the Feeble Mind: A History of Mental Retardation in the United States.* Berkeley and Los Angeles: University of California Press.

Trombley, Stephen. 1988. *The Right to Reproduce: A History of Coercive Sterilization.* London: Weidenfeld and Nicolson.

Tucker, William H. 1994. *The Science and Politics of Racial Research.* Urbana: University of Illinois Press.

Tuttle, Florence Guertin. 1921. "Why I Believe in Birth Control." *Birth Control Review* 5:5–7.

Twiss, Seymour B. 1979. "The Genetic Counselor as Moral Advisor." *Birth Defects Original Article Series* 15:201– 12.

van Wagenen, Bleeker. [1912] 1984. "Preliminary Report to the First International Eugenics Congress of the Committee of the Eugenics Section, American Breeders' Association to Study and Report as to the Best Practical Means for Cutting off the Defective Germ-Plasm in the Human Population." In *Problems in Eugenics: Papers Communicated to the First Interna-*

tional Eugenics Congress. Reprint. New York: Garland.

Wallace, Alfred Russel. [1864] 1991. "The Origin of Human Races and the Antiquity of Man Deduced from the Theory of 'Natural Selection.'" In *Alfred Russel Wallace: An Anthology of His Shorter Writings*, ed. C. H. Smith. New York: Oxford University Press.

Wallace, Alfred Russel. 1870. "The Limits of Natural Selection as Applied to Man." In *Contributions to the Theory of Natural Selection: A Series of Essays.* London: Macmillan.

Wallace, Alfred Russel. [1890] 1991. "Human Selection." In *Alfred Russel Wallace: An Anthology of His Shorter Writings*, ed. C. H. Smith. Oxford: Oxford University Press.

Wallace, Alfred Russel. 1900. *Studies: Scientific and Social.* London: Macmillan.

Wallace, Alfred Russel. 1905. *My Life: A Record of Events and Opinions.* Vol. 2. London: Chapman and Hall.

Wallace, Alfred Russel. [1912] 1991. Interview fragment. In *Alfred Russel Wallace: An Anthology of His Shorter Writings*, ed. C. H. Smith. Oxford: Oxford University Press.

Wallace, Alfred Russel. [1913] 1991. "Social Environment and Moral Progress." In *Alfred Russel Wallace: An Anthology of His Shorter Writings*, ed. C. H. Smith. New York: Oxford University Press.

Walter, Herbert Eugene. 1913. *Genetics: An Introduction.* New York: Macmillan.

Ward, Lester Frank. 1891. "The Transmission of Culture." In *Glimpses of the Cosmos: A Mental Autobiography*, vol. 4. New York: Putnam.

Webb, Sidney. 1910–11. "Eugenics and the Poor Law: The Minority Report." *Eugenics Review* 2:233–41.

Weindling, Paul. 1989. *Health, Race, and German Politics between National Unification and Nazism, 1870–1945.* Cambridge: Cambridge University Press.

Weiss, Sheila Faith. 1987. *Race, Hygiene, and National Efficiency: The Eugenics of Wilhelm Schallmayer.* Berkeley and Los Angeles: University of California Press.

Weiss, Sheila Faith. 1990. "The Race Hygiene Movement in Germany, 1904–1945." In *The Wellborn Science: Eugenics in Germany, France, Brazil, and Russia*, ed. M. Adams. New York: Oxford University Press.

Wells, H. G. 1905. Comment. In *Sociological Papers*, vol 1. London: Macmillan.

Wells, H. G. [1922] 1950. Preface to M. Sanger, *The Pivot of Civilization.* Reprint. Elmsford, N.Y.: Maxwell Reprint.

Werskey, Gary. 1978. *The Visible College: A Collective Biography of British Scientific Socialists of the 1930s.* New York: Holt, Rinehart, and Winston.

Wertz, Dorothy C., and John C. Fletcher. 1989. "Ethical Decision Making in Medical Genetics: Women as Patients and Practitioners in Eighteen Nations." In *Healing Technology: Feminist Perspectives*, ed. K. Ratcliff. Ann Arbor: University of Michigan Press.

Wheeler, J. H. 1885. "Heredity and Progress." *Progress* 5 (November): 499.

Wiggam, Albert E. 1924. *The Fruit of the Family Tree.* Indianapolis: Bobbs-Merrill.

Wolfensberger, Wolf. 1975. *The Origin and Nature of Our Institutional Models.* Syracuse: Human Policy Press.

Woods, Frederick Adams. 1913. "Heredity and the Hall of Fame." *Popular Science Monthly* 82 (May): 445–52.

Woods, Frederick Adams. 1918. Review. *Science* 48 (October):419–20.

Woods, Frederick Adams. 1923. "A Review of Reviews of Madison Grant's *Passing of the Great Race." Journal of Heredity* 14 (May): 93–95.

Wright, Robert. 1990. "Achilles' Helix." *New Republic* (July 9, 16).

Yerkes, Robert M., ed. 1921. *Psychological Examining in the United States Army.* Vol. 15 of *Memoirs of the National Academy of Sciences.* Washington, D.C.

Zenderland, Leila. 1992. "A Sermon of New Science: *The Kallikak Family* as Eugenic Parable." Paper presented to the History of Science Society, Washington, D.C.

Zenderland, Leila. Forthcoming. *Measuring Minds: Henry Herbert Goddard and the Origins of American Intelligence Testing.* New York: Cambridge University Press.

Index

151

DISCARD

DATE DUE